FLUID MECHANICS

Fluid Mechanics

A Geometrical Point of View

S. G. Rajeev

Department of Physics and Astronomy
Department of Mathematics
University of Rochester
Rochester, NY

OXFORD

UNIVERSITY PRESS

OXFORD
UNIVERSITY PRESS

Great Clarendon Street, Oxford, OX2 6DP,
United Kingdom

Oxford University Press is a department of the University of Oxford.
It furthers the University's objective of excellence in research, scholarship,
and education by publishing worldwide. Oxford is a registered trade mark of
Oxford University Press in the UK and in certain other countries

First Edition published in 2018
Impression: 1

Published in the United States of America by Oxford University Press
198 Madison Avenue, New York, NY 10016, United States of America

British Library Cataloguing in Publication Data
Data available

Library of Congress Control Number: 2018932827

ISBN 978-0-19-880502-1 (hbk.)
ISBN 978-0-19-880503-8 (pbk.)

DOI: 10.1093/oso/9780198805021.001.0001

Printed and bound by
CPI Group (UK) Ltd, Croydon, CR0 4YY

This book is dedicated to my teacher, A. P. Balachandran.

Preface

Next to celestial mechanics, fluid mechanics is the oldest part of theoretical physics. Euler derived the fundamental equations more than 200 years ago. Yet it is far from complete. We do not yet know with mathematical certainty if Euler's equations have regular solutions given smooth initial data. More important to physics, the phenomenon of turbulence is still mysterious.

Numerical methods have made much progress (especially in applications to engineering) in recent years. The exponential growth of computing power has made it possible to design airplanes, submarines and household appliances without cumbersome testing of prototypes. Local weather can be predicted for about a fortnight, after which even the best computers fail. Tools from statistical mechanics and quantum field theory (the areas that routinely deal with an infinite number of random variables) ought to be useful.

Many ideas of theoretical physics (e.g., conformal invariance) originated in the study of fluids. Abstract ideas (such as Lie algebras) appear here in a concrete and easily visualized setting. In this book I want to present fluid mechanics as a nonlinear classical field theory, an essential part of the education of a physicist.

The central object of fluid mechanics is a vector field, the fluid velocity. So, a geometric point of view is quite natural. At a deeper level, one can understand Euler's equation as a hamiltonian system, whose Poisson brackets are dual to the commutation relations of vector fields and the hamiltonian is the kinetic energy (L^2-norm).

This allows one to think of Euler's equations for fluids and his equation for a rigid body on the same footing: they both describe geodesics on a Lie group. Of course, the fluid is much more complicated: the group is infinite dimensional and the curvature is negative. Arnold showed that the notorious instabilities of fluid mechanics can be understood in terms of the negative curvature of its geometry. Although very natural, this needs some mathematical machinery. I have tried to present it in a way accessible to theoretical physicists. Clebsch's old idea of using canonical variables in fluid mechanics becomes useful here.

More traditional subjects like boundary layer theory, vortex dynamics, and surface waves are also illuminated by geometry. Expressing Euler and Navier–Stokes equations on curvilinear coordinates is essential: most boundaries of interest are not planes. A little Riemannian geometry goes a long way here. The linear theory of instabilities is understood in terms of the spectrum of non-normal operators. Hermann Weyl (of all people) found an excellent analytical approximation for the boundary condition–Blasius theory of boundary layers. The vortex filament equations can be related to the Heisenberg model of magnets; a vortex in fluid mechanics is mapped to a soliton of spin waves.

Throughout, I will point to numerical and analytical calculations (mostly using Mathematica, but you can use your favorite language) to gain understanding. This is not a book on computational methods. You don't need a supercomputer anymore to do interesting work, so even the most theoretical of us can benefit from their use. Switching from an analytical/geometrical point of view to the more practical discrete/numerical one and back, we understand both better.

Studying chaotic advection is a good way of developing intuition for fluid flows and for dynamical systems. Aref (1984) gives a simple example, which has turned out to be useful in engineering devices that mix fluids efficiently. The Smale horse shoe, although not usually thought of as part of fluid mechanics, gives a solidly established mathematical model of chaos. Many chaotic systems (including Aref's) have a Smale horse shoe embedded inside them. So I have included a discussion of these topics in an appendix.

There are many reasons to believe that renormalization is useful to understand turbulence. Although a discussion of those theories is beyond the scope of this book, I have included an appendix on more elementary applications of renormalization: the Ising model and Feigenbaum's approach to dynamical systems in one dimension.

I have not tried to survey every sub-field in detail: such a comprehensive survey would be about as useful as a map of a country on a scale of 1:1. A few cases are examined in detail. Theoretical physics is based on general principles (conservation laws, variational principles) and special cases that can be understood analytically or by simple numerical computations. The techniques developed this way can be adapted to those that arise in applications. That said, my choice of topics is necessarily subjective. The emphasis on Lie theory and dynamical systems is unusual for a book at this level.

Much remains uncovered. It would have been nice to talk of quantum fluids. The great mystery of turbulence and attempts to model it could have been reviewed. Models of weather prediction, oceanography, and astrophysical jets are all thriving. The whole field of magneto hydrodynamics is given short shrift. Each of these topics would require a book of its own.

This is a sort of sequel to *Advanced Mechanics* Rajeev (2013). A knowledge of mechanics, linear algebra (eigenvalue problems) and partial differential calculus is the main pre requisite. The sections with a ∗ in their name can be skipped on a first reading; they contain more advanced material.

The book is aimed at physics/mathematics graduate students; some engineering students will also find it useful. There is some overlap and some repetition, because some ideas (e.g., dynamical systems, Lie algebras) are so recurrent in physics. Others are only outlined, and working them out yourself is an essential part of learning the subject.

Acknowledgments

I cannot thank my wife and children enough for their support over the years. A. P. Balachandran taught much more than physics: the audacity to go into new areas. Thanks to the indulgence of my colleagues in Physics and Mathematics Departments for letting me explore directions that are not profitable in the short term.

I am grateful to A. Kar, G. Krishnaswami and M. Bhattacharya for commenting on parts of the manuscript. Thanks for discussions with A. Iosevich on fractals; D. Geba on regularity of Navier–Stokes; V. V. Sreedhar on conformal invariance; and V. P. Nair and T. Padmanabhan on countless topics of physics in our early years. Thanks to Sonke Adlung for encouraging me to write this book and to Harriet Konishi for many helpful suggestions. Thanks also to Alan Skull and Lydia Shinoj for the excellent editing.

Contents

List of figures

1

Vector Fields

A fluid is composed of a large number of molecules. These molecules are in rapid motion, each in a different direction. They collide with each other, which tends to randomize the molecular velocities. In the fluid approximation, we think of the system as composed of "fluid elements" which are large enough to contain a multitude of molecules but still small compared to the size of the vessel containing them.

The average velocity of a fluid element will be much smaller than the individual molecular velocities. The distance between molecules is so small that we can regard the density and velocity of a fluid as continuous functions of space and time. A function in space-time is called a field. Thus, the fluid pressure is a scalar field while fluid velocity is a vector field. Other examples of scalar fields are temperature, entropy, and the concentration of some chemical pollutant carried by the fluid.

1.1 The velocity field

Let x^i for $i = 1, 2, 3$ be the coordinates of some point within the fluid. To begin with, think of them as Cartesian coordinates. Later, we will see that curvilinear co-ordinates work just as well. We will follow the convention of geometry in writing the index on coordinates as superscripts. $v^i(x, t)$ are the components of the fluid velocity at time t and position x (sometimes we will omit the index on the coordinate). This means that a fluid element at x^i at time t will move to $x^i + \epsilon v^i(x, t)$ at the next instant $t + \epsilon$, where ϵ is an infinitesimally small time interval.

Given a scalar field $f(x, t)$, there are two notions of time derivative that are important in fluid mechanics. The obvious one is the partial derivative,

$$\frac{\partial f}{\partial t}(x, t) = \lim_{\epsilon \to 0} \frac{f(x, t + \epsilon) - f(x, t)}{\epsilon},$$

in which we look at the change in time at a fixed location in the fluid element. The other is the total derivative or material derivative

$$\frac{Df}{dt} = \frac{\partial f}{\partial t} + v^i \frac{\partial f}{\partial x^i}.$$

Fluid Mechanics: A Geometrical Point of View. S. G. Rajeev © S. G. Rajeev 2018.
Published in 2018 by Oxford University Press. DOI: 10.1093/oso/9780198805021.001.0001

1.2   Space-time approach

In relativistic physics, we must think of physical quantities as functions of space-time. This is a four-dimensional manifold whose coordinates x^μ are the three coordinates of space x^i, $i = 1, 2, 3$ plus time t, which is usually thought of as the zeroth coordinate of space-time $x^0 = t$. This point of view is also convenient in some non-relativistic situations,[1] including fluid mechanics. If we set $v^0 = 1$:

$$\frac{\partial f(t, x)}{\partial t} + v^i(t, x)\frac{\partial f(t, x)}{\partial x^i} \equiv v^\mu(x)\frac{\partial f}{\partial x^\mu}.$$

$$\mu = 0, 1, 2, 3, \quad i = 1, 2, 3$$

Thus the velocity field can be thought as a first order differential operator (i.e., the material derivative) in space-time:

$$v = \frac{\partial}{\partial t} + v^i(t, x)\frac{\partial}{\partial x^i}.$$

Even if the fluid elements are at rest, $v^i = 0$, they are moving forward in time. Even if the fluid is moving, the rate at which it moves forward in time is unaffected. (This is the non-relativistic approximation). The coordinates (t, x) do not need to be Cartesian; and they do not need to be measured in an inertial frame. We can make any (possibly time-dependent) transformation of the space coordinates. But the time coordinate should not change, because in the non-relativistic limit all observers agree on the definition of time. Of course, there is a theory of relativistic fluids (Poisson and Will, 2014). This book happens not to be about that.

$$\tilde{t} = t, \quad \tilde{x}^i = \phi^i(t, x),$$

$$vf = \frac{\partial f}{\partial t} + \tilde{v}^i(t, x)\frac{\partial f}{\partial \tilde{x}^i} = \frac{\partial f}{\partial t} + v^i(t, x)\frac{\partial f}{\partial x^i}.$$

Applying the chain rule of differentiation

$$\tilde{v}^\mu(x) = \frac{\partial \phi^\mu(x)}{\partial x^\nu}v^\nu(x).$$

Since $\frac{\partial \phi^0(t,x)}{\partial t} = 1$, $\frac{\partial \phi^0(t,x)}{\partial x^i} = 0$,

$$\tilde{v}^0(t, x) = 1,$$

---

[1] Recall the title of Feynman's classic paper, *Space-Time Approach to Non-Relativistic Quantum Mechanics*.

and

$$\tilde{v}^i(t,x) = \frac{\partial \phi^i(t,x)}{\partial t} + v^j(t,x)\frac{\partial \phi^i(t,x)}{\partial x^j}. \tag{1.1}$$

Thus time-dependent transformations of the type we are considering preserve the condition that the time component of velocity is one.

1.3 Eulerian vs Lagrangian picture

In non-relativistic mechanics there is something special about an inertial reference frame: Newton's laws as originally stated hold in such frames. In fluid mechanics this is called the Euler frame or the Eulerian picture. A transformation from one Eulerian frame to another is time-independent. It can still transform space coordinates in any (possibly nonlinear) way.

If we allow time-dependent transformations, we can even choose ϕ^i such that the fluid velocity is zero everywhere in the new system. The new system would be co-moving with the fluid: a non-inertial reference frame. This is the Lagrangian picture. The Eulerian and Lagrangian pictures are related by the transformation satisfying the differential equation obtained by setting $\tilde{v}^i = 0$ in eqn (1.1):

$$\frac{\partial \phi^i(t,x)}{\partial t} + v^j(t,x)\frac{\partial \phi^i(t,x)}{\partial x^j} = 0. \tag{1.2}$$

We will mostly stick to the Eulerian picture, as dynamical laws of fluid motion (such as the Euler equations) are easiest to understand from this point of view. But some persistent structures of fluids (vortices, jet streams) are easier to see in the Lagrangian picture. Sometimes we will switch to this picture.

1.4 Integral curves

Imagine a speck of dust carried along (or advected) by the fluid. Its position $\xi^i(t)$ will change with time according to the differential equation

$$\frac{d\xi^i(t)}{dt} = v^i(\xi(t), t).$$

Given its initial position, $\xi^i(0)$, this differential equation can be solved to determine the path followed by the dust particle. Geometrically, this is a curve starting at $\xi^i(0)$ whose tangent at each point is the velocity vector at that point (and instant of time). The process of solving a differential equation is a kind of integration. So $\xi^i(t)$ is called the integral curve of the vector field $v^i(x,t)$. Exactly one such curve passes through every point in space. Piecing together all the integral curves, we have a function $\Xi^i(x,t)$ satisfying

$$\frac{\partial \Xi^i(x,t)}{\partial t} = v^i\left(\Xi(x,t), t\right)$$

and the initial condition

$$\Xi^i(x,0) = x^i.$$

1.5   The method of characteristics

A scalar field is said to be advected by the flow $v^i(x,t)$ if it satisfies

$$\frac{Df}{dt} \equiv \frac{\partial f}{\partial t} + v^i(x,t)\frac{\partial f}{\partial x^i} = 0.$$

That is, it is constant along the path of a particle carried along by the flow. Suppose we are given the initial value of this scalar field $f_0(x) = f(x,0)$ and we want to predict what will be its value at some later time t. We must then start at the point x at time t and trace back along the integral curve of v to find the point $\eta(x,t)$ where it was at time zero. You can see that this is the inverse of the problem we discussed in the section on integral curves:

$$\Xi(\eta(x,t), t) = x.$$

So if we know all the integral curves $\Xi(x,t)$ and can find the inverse function above, we can solve the equation for advected scalar fields

$$f(x,t) = f_0(\eta(x,t)).$$

Since $\Xi(x,t)$ is determined by $v(x,t)$, so is its inverse $\eta(x,t)$. For example, if $v(x,t) = v$ a constant, the solution is $f(x,t) = f_0(x - vt)$ as can be easily verified.

This method of using a system of ordinary differential equations (ODEs) to solve a partial differential equation (PDE) is a particular case of the method of characteristics. See (Courant and Hilbert, 1962), Chapter 1 for an exposition of the general case.

When the vector field is time-independent (steady) we can expect a solution to the advection equation that is also steady. That is,

$$v^i\frac{\partial f}{\partial x^i} = 0.$$

Example 1.1

If $v(x) = \left(-x^2, x^1\right)$ the integral curves are circles. A steady solution to the advection equation is any function that is rotation invariant.

The paths of advected particles can be quite complicated; in general they can be determined only by numerical integration.

Example 1.2 The double gyre

$$v_x = \partial_y A, \quad v_y = -\partial_x A, \quad A = -\alpha \sin\left[\pi\theta(x, t)\right] \sin \pi y,$$

where

$$\theta(x, t) = a(t)x^2 + b(t)x,$$
$$a(t) = \epsilon \sin t, \quad b(t) = 1 - 2\epsilon \sin t.$$

The parameter ϵ controls the time dependence; if it is zero the velocity is independent of time.

Exercise 1.1 Write a program in Mathematica (or your own favorite language) which plots the integral curves of the double gyre for various values of ϵ and the initial point. A small change in the initial condition can lead to a big change in the outcome after enough time. An animation will be more interesting than a static plot. The values $\alpha = 1.1, \epsilon = 0.25$ are good choices.

1.6   Conservation law

The total mass contained within some region V of the fluid is $\int_V \rho(x, t)dx$. Since mass is conserved in non-relativistic physics, the change of this quantity must be equal to the inflow of mass into the region through its boundary.

$$\frac{\partial}{\partial t} \int_V \rho(x, t)dx = -\int_{\partial V} \rho(x, t)v^i(x, t)dS_i.$$

Here dS_i is the area of an infinitesimal element on the boundary, which is thought of as a vector pointing along the outward normal. Appealing to Gauss' theorem

$$\int_{\partial V} \rho(x,t)v^i(x,t)dS_i = \int_V \frac{\partial \left[\rho v^i\right]}{\partial x^i} dx.$$

Thus the conservation of mass becomes (sometimes we will suppress the arguments of functions for simplicity of notation)

$$\int_V \left\{ \frac{\partial \rho}{\partial t} + \frac{\partial \left[\rho v^i\right]}{\partial x^i} \right\} dx = 0.$$

Since this holds for any region V we must have

$$\frac{\partial \rho}{\partial t} + \frac{\partial \left[\rho v^i\right]}{\partial x^i} = 0.$$

The same argument holds for any scalar quantity that is conserved. For example, suppose our fluid is a solution of two kind of molecules that do not interact chemically with each other. Then their number densities $\rho_1(x,t)$, $\rho_2(x,t)$ will be separately conserved.

$$\frac{\partial \rho_1}{\partial t} + \frac{\partial \left[\rho_1 v^i\right]}{\partial x^i} = 0, \quad \frac{\partial \rho_2}{\partial t} + \frac{\partial \left[\rho_2 v^i\right]}{\partial x^i} = 0.$$

Suppose $\phi(x,t) = \frac{\rho_1(x,t)}{\rho_2(x,t)}$ is the relative concentration of one type of molecule compared to the other. From the conservation laws we can deduce that

$$\frac{\partial \phi}{\partial t} + v^i \frac{\partial \phi}{\partial x^i} = 0.$$

That is, the relative concentration is an advected scalar field. This is one of the reasons why advected scalars are interesting in fluid mechanics.

1.7   Densities vs scalars

Not all physical quantities described by a single real number at each point are scalars. For example, they could be densities. The difference is in how they change under a coordinate transformation. If we make a change of coordinates (assumed to be time-independent for simplicity)

$$\tilde{x}^i = \phi^i(x),$$

a scalar transforms in a simple way:

$$\tilde{f}(\tilde{x}) = f\left(\phi^{-1}(\tilde{x})\right),$$

where ϕ^{-1} is the inverse transformation.

$$\phi\left(\phi^{-1}(x)\right) = x.$$

If the transformation is infinitesimal (i.e., differs from the identity by an infinitesimally small quantity times a vector field),

$$\phi^i(x) \approx x^i + \epsilon v^i(x), \quad |\epsilon| << 1$$
$$x^i \approx \tilde{x}^i - \epsilon v^i(x).$$

the change in the function is also small

$$\tilde{f}(\tilde{x}) \approx f(x) - \epsilon v^i \partial_i f.$$

Note that the change is (up to a factor $-\epsilon$) the directional derivative of the function along a vector field whose components are v^i. Vector fields are infinitesimal transformations of space. The change of any quantity under the action of a vector field is called its Lie derivative. For a scalar it is just the directional derivative:

$$\mathcal{L}_v f = v^i \partial_i f.$$

A density describes the amount of some physical quantity (mass, charge, etc.) in a small volume. So it is the combination $\rho(x) d^3 x$ which is invariant under coordinate transformations.

$$\tilde{\rho} d^3 \tilde{x} = \rho d^3 x.$$

Recall that

$$d^3\tilde{x} = \det \frac{\partial \phi}{\partial x} d^3 x, \quad \frac{\partial \phi}{\partial x} = \begin{pmatrix} \frac{\partial \phi^1}{\partial x^1} & \cdots & \frac{\partial \phi^1}{\partial x^3} \\ \cdots & \cdots & \cdots \\ \frac{\partial \phi^3}{\partial x^1} & \cdots & \frac{\partial \phi^3}{\partial x^3} \end{pmatrix}.$$

Thus, $\tilde{\rho}(\tilde{x}) = \det \frac{\partial \phi^{-1}}{\partial \tilde{x}} \rho\left(\phi^{-1}(\tilde{x})\right)$. Infinitesimally,

$$\frac{\partial \phi^{-1}}{\partial \tilde{x}} \approx 1 - \epsilon \frac{\partial v}{\partial x}, \quad \det \frac{\partial \phi^{-1}}{\partial \tilde{x}} \approx 1 - \epsilon \mathrm{tr} \frac{\partial v}{\partial x} \approx 1 - \epsilon \partial_i v^i.$$
$$\tilde{\rho}(\tilde{x}) \approx \rho(x) - \epsilon v^i \partial_i \rho - \epsilon \left[\partial_i v^i\right] \rho(x) = \rho(x) - \epsilon \partial_i \left[\rho v^i\right].$$

Thus, the infinitesimal change of a density under a vector field is the coefficient of $-\epsilon$ in the above

$$\mathcal{L}_v \rho = \partial_i \left[\rho v^i\right].$$

For a scalar on the other hand,

$$\mathcal{L}_v \phi = v^i \partial_i \phi.$$

1.8   Steady flows

A vector field whose components in an inertial frame are time-independent is said to be a steady flow. Such flows are the analog of static configurations in mechanics. Even though the individual fluid elements are moving, the fluid as a whole is in equilibrium.

1.9   Incompressible flow

A vector field whose divergence is zero

$$\nabla \cdot v = 0$$

is said to be *incompressible*. This is just the statement of the conservation of mass when the density ρ is a constant.

If there is a vector field A such that

$$v = \nabla \times A,$$

then the divergence is zero. In the special case of two-dimensional incompressible flow on a plane, this vector field points in the direction normal to the plane. So it may be considered to be a scalar function on the plane:

$$v_x = \partial_y A, \quad v_y = -\partial_x A.$$

It is called the *stream function*.

1.10   Irrotational flow

If the vorticity

$$\omega = \nabla \times v$$

is zero in the domain of interest, we say that the flow is irrotational. If there is a scalar field such that the vector field is its gradient

$$v = \nabla \phi$$

then it is irrotational:

$$\nabla \times v = 0.$$

This follows from the definition of the curl. ϕ is called the velocity potential. Such a flow is also called a gradient flow.

Along the integral curves of a gradient flow,[2] the velocity potential is monotonically increasing

$$\frac{D\phi}{dt} = v^i \partial_i \phi = \partial_i \phi \partial_i \phi \geq 0.$$

This vanishes only at the fixed points, where the vector field is zero. These fixed points are extrema of the potential. At any point, the gradient vector field points in the direction in which ϕ increases the fastest.

The converse is not always true: not all irrotational vector fields are gradients. Consider the flow in the plane

$$v_x = -\frac{x}{x^2 + y^2}, \quad v_y = \frac{y}{x^2 + y^2}.$$

in the annular region with $x^2 + y^2 > a^2$ for some constant a. It is irrotational. Yet the function of which it is the gradient is multi-valued within the domain of interest:

$$v = \nabla \arctan \frac{y}{x}.$$

In fact $\arctan \frac{y}{x} = \theta$, the angle of the polar coordinate system $x = r\cos\theta, y = r\sin\theta$. Each time we go around a closed curve surrounding the bounding circle, $\log\theta$ jumps by 2π. This is a minor point in most physical applications, but does lead to some interesting situations in quantum fluid flow: the vorticity has to be quantized in multiples of $\hbar$.

1.11 Irrotational and incompressible flow

A vector field is incompressible if it has zero divergence

$$\nabla \cdot v = 0.$$

If a vector field is both irrotational and incompressible, the potential must satisfy the Laplace equation:

$$v = \nabla \phi, \quad \nabla \cdot \nabla \phi = 0.$$

In the early days of fluid mechanics, much effort was expended understanding solutions to this equation for various boundary conditions. Even now it is a good starting point. We will return to this.

---

[2] We consider for now only time-independent flows.

1.12   Jacobi matrix

The Jacobi matrix $\mathcal{F}_j^i(x, t) = \frac{\partial \phi_t^i(x)}{\partial x^j}$ describes the change in outcome of a small change in initial point. It is not difficult to derive a differential equation for it:

$$\frac{\partial \mathcal{F}_j^i(x, t)}{\partial t} = \frac{\partial v^i}{\partial x^j}(\phi_t(x), t), \quad \mathcal{F}_j^i(x, 0) = \delta_j^i.$$

....

Exercise 1.2 Show that if the vector field has zero divergence, $\partial_i v^i = 0$ the Jacobi matrix has determinant one. (It is useful to remember that $\log \det M = \mathrm{tr}\ \log M$ and therefore that $\delta \det M = \mathrm{tr}\ M^{-1} \delta M$.)

....

Let us look at some examples.

Example 1.3

If the fluid is at rest $v^i(x, t) = 0$, advection is trivial: the position of an advected particle is simply constant $\xi^i(t) = \xi^i(0)$. The material and partial derivatives w.r.t. time are the same.

Example 1.4

The next simplest case is that of a fluid moving in some constant direction (say the first direction) at constant speed v. The integral curves are straight lines

$$\xi^1(t) = \xi^1(0) + vt, \quad \xi^i(t) = \xi^i(0), \quad i = 2, 3.$$

Example 1.5

If the fluid is rotating at a constant rate around some point we have a vortex.

$$v_x = -\frac{1}{2}\omega y, \quad v_y = \frac{1}{2}\omega x, \quad v_z = 0.$$

The center of a vertex is a *fixed point*: an advected particle starting at that point remains there for ever. But particles starting at other points are in uniform circular motion; the integral curves are circles.

Exercise 1.3 Determine the integral curves of the vector field $v^i(x, t) = x^i$. Plot the curves.

1.13 Flow near a fixed point

A point at which a steady flow vanishes is called a *fixed point*. A fluid element located at a fixed point remains there for ever. In the immediate vicinity of a fixed point, we can approximate the flow by expanding in a Taylor series. This gives us a matrix of derivatives of the components. The eigenvalues and eigenvectors of this matrix give a complete description of the flow in a linear approximation. Let us choose the origin at the fixed point. Then $v^i \approx \mathcal{J}^i_j x^j$, where $\mathcal{J}$ is the derivative evaluated at the origin (the *infinitesimal Jacobi matrix*) $\mathcal{J}^i_j = \frac{\partial v^i}{\partial x^j}$.

The equation for integral curves is linear in this approximation:

$$\frac{d\xi^i}{dt} = \mathcal{J}^i_j \xi^j.$$

or

$$\frac{d\xi}{dt} = \mathcal{J}\xi$$

in matrix notation. Let us consider the various types of flow near this fixed point. The situation is easier to visualize in two dimensions so we begin there.

Example 1.6 Stable fixed point

$$v^1(x, t) = -x^1, \quad v^2 = -x^2.$$

The integral curves are $\xi^1(t) = X^1 e^{-t}, \xi^2(t) = X^2 e^{-t}$. They start at (X^1, X^2) and move towards the origin.

Example 1.7 Parabolic fixed point

The opposite situation arises with

$$v^1(x, t) = x^1, \quad v^2 = 2x^2$$

with the integral curves pointing outward along some parabolas.

Example 1.8 Hyperbolic fixed point

The simplest example is $v^1 = x^1$, $v^2 = -x^2$. The unstable manifold is the x^1-axis and the stable manifold is the x^2-axis. The integral curves are hyperbolas with these as axes.

Exercise 1.4 Consider vector field $v^1(x, t) = \alpha x^1 + \beta x^2$, $v^2 = \gamma x^1 + \delta x^2$, where $\alpha, \beta, \gamma, \delta$ are real numbers such that $D = \alpha\delta - \beta\gamma < 0$. Show that the eigenvalues of the infinitesimal Jacobi matrix are real and that one is positive and one negative. Find the stable and unstable directions for the values $\alpha = 0.75$, $\beta = 0.7$, $\gamma = 0.9$, $\delta = -0.71$. Plot some integral curves of vector fields near the fixed point.

Example 1.9 Three dimensions

A fixed point which is stable in the first two directions and unstable in the third is given by

$$
v = \begin{pmatrix} \alpha & \beta & 0 \\ -\beta & \alpha & 0 \\ 0 & 0 & -2\alpha \end{pmatrix} \begin{pmatrix} x^1 \\ x^2 \\ x^3 \end{pmatrix}, \quad \alpha = -0.05, \quad \beta = -0.04.
$$

Points close to the $\left(x^1, x^2\right)$ plane will circulate around the origin while converging to it, and are then ejected along the third axis.

For topological reasons, every vector field on a sphere must vanish somewhere. Here is an example of a vector field that has no fixed points.

Example 1.10 Foliation of a torus

A foliation is a vector field that is non-zero everywhere. A torus is the only compact manifold of dimension two on which a foliation can exist. The simplest case is a vector field with constant components; the integral curves are straight lines. Locally (within a small enough neighborhood) any foliation can be reduced to this case by a change of coordinates. The global behavior can be quite intricate.

Exercise 1.5 Often we impose periodic boundary condition on a rectangular region for theoretical convenience. This is the same as a torus, the doughnut shaped surface you get by gluing the opposite sides of a square to each other.

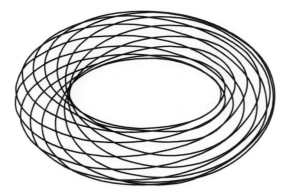

Figure 1.1 *The integral curve of the constant vector field* $\left(1, \frac{8}{13}\right)$ *on the torus*

1. Visualize this, using an explicit function that embeds the unit square with periodic boundary conditions into $\mathbb{R}^3$:

$$(x, y) \mapsto (\cos[2\pi x]\{a + \cos[2\pi y]\}, \sin[2\pi x]\{a + \cos[2\pi y]\}, \sin 2\pi y).$$

Plot this surface for some choices of the parameter a.

2. Show that the integral curves of a constant vector field (v^1, v^2) are locally, straight lines in the square $\xi(t) = (x^1 + v^1 t, x^2 + v^2 t)$. What happens to them at the boundary?

3. When the slope $\omega = \frac{v^2}{v^1}$ is rational $\left(e.g., \frac{8}{13}\right)$ the integral curve is closed. Plot this both on the square and on the embedded torus (Figure 1.1).

4. Show that, as the slope varies over a sequence of rationals that tend to an irrational number, the integral curves fill out more and more of the torus. In the limit we get a space-filling curve: it comes as close to any point as you want if you extend it far enough. Plot the integral curves for the sequence of continued fractions that tend to the Golden Ratio: $\omega = 1 + \cfrac{1}{1 + \cfrac{1}{1 + \cdots}}$.

Exercise 1.6 Solve numerically the Lorenz system

$$\frac{dx}{dt} = \alpha(y - x), \quad \frac{dy}{dt} = -xz + \beta x - y, \quad \frac{dz}{dt} = xy - z,$$

and plot the solution in three dimensions. For example, set $\alpha = 3, \beta = 20.5$. Show that after a long time (e.g., $T = 150$) any initial point will end up in a subset ("attractor") shaped like the wings of a butterfly. Yet, it is impossible to tell which wing the point will be on, as even the tiniest change in initial conditions will kick it from one wing to the other.

2

Euler's Equations

There is a lot going on in this chapter. The best strategy is to read through and return to sections as needed later on in the book.

2.1 Conservation of momentum

The same idea we used to derive conservation of mass can be used to derive an equation for the conservation of momentum. We consider a small fluid element and ask how momentum flows into it. The change in momentum must be equal to the force, which in an ideal fluid is the negative gradient of pressure plus whatever external force (e.g., gravity) is acting on the fluid. Ideal means that we ignore dissipative forces such as friction (with a boundary) or viscosity. We will see how to include these forces later.

The quantity ρv^i is the density of the ith Cartesian component of momentum. The ith component of the total momentum of the fluid in some region V is

$$\int_V \rho v^i d^3 x.$$

For each i, the amount of momentum flowing out of a region V is

$$\int_{\partial V} \left[\rho v^i\right] v \cdot dS = \int_{\partial V} \rho v^i v^j dS_j.$$

Recall that dS^i is the area of a small surface element, thought of as a vector pointing outward in the normal direction.

By applying Gauss' theorem to each component i this can be rewritten as a volume integral

$$\int_{\partial V} \rho v^i v^j dS_j = \int_V \partial_j \left[\rho v^i v^j\right] d^3 x.$$

Fluid Mechanics: A Geometrical Point of View. S. G. Rajeev © S. G. Rajeev 2018.
Published in 2018 by Oxford University Press. DOI: 10.1093/oso/9780198805021.001.0001

Thus the total change in the momentum of a region is

$$\frac{\partial}{\partial t} \int_V \rho v^i d^3 x + \int_V \partial_j \left[\rho v^i v^j \right] d^3 x.$$

This must be equal to the net force acting on this region of fluid.

The most important such force is due to the pressure differential at different points in the fluid. Again, if you consider a small cube at x, the force acting on the face normal to the first axis at x is $p dx^2 dx^3$ (i.e., pressure times area). That at the opposite face is $[p + \partial_1 p dx^1] dx^2 dx^3$. Thus Newton's second law gives

$$\frac{\partial \left[\rho v^1 \right]}{\partial t} + \partial_j \left[\rho v^1 v^j \right] = -\partial_1 p + f_1,$$

where f_1 is the first component of the net force from all sources other than pressure. The negative sign takes into account that the force due to pressure is thought of as pointing outward. Taking into account all directions,[1]

$$\frac{\partial \left[\rho v^i \right]}{\partial t} + \partial_j \left[\rho v^i v^j \right] = -\partial_i p + f_i. \tag{2.1}$$

This is *Euler's equation.*

Here f_i includes gravity, friction, viscosity, etc. For an ideal fluid (ignoring friction, viscosity, and other dissipative forces) in the absence of external forces we get

$$\frac{\partial \left[\rho v^i \right]}{\partial t} + \partial_j \left[\rho v^i v^j \right] = -\partial_i p.$$

So we now have five unknowns ρ, v^i, p at each point in space-time, and only four equations, for the conservation of mass and momentum. The remaining piece of the puzzle is an equation of state that determines p from ρ. This depends on the nature of the fluid. We will first get an overview of the various special cases of interest, and return to some of them for detailed study later.

2.2 The stress tensor

In the absence of external forces, Euler's equation can be written as a conservation law

$$\frac{\partial \left[\rho v^i \right]}{\partial t} + \partial_j T^{ij} = 0, \quad T^{ij} = \rho v^i v^j + p \delta^{ij}.$$

---

[1] We use the Euclidean metric to raise and lower indices. It is therefore not necessary to distinguish between upper and lower indices. But we keep the conventions of geometry intact to some extent, for clarity.

The first term represents the change of momentum in a small volume element of the fluid. The second term represents the flow of momentum out of this region. If the conserved quantity were a scalar (like mass or charge) the current associated with it would have been a vector j^i, describing how much of it is flowing across a surface whose normal is pointing in the ith direction. Momentum being a vector, we must think of the flow of each of its components separately. So we get a matrix (also called tensor) T^{ij} which measures how much of the ith component of momentum flows across a surface whose normal is oriented in the jth direction. Since the change of momentum is force, this can also be thought of as the force per unit area: the force and area both being vectors, we need a matrix to represent their ratio. "Stress" is an old engineering term for force per unit area, hence T^{ij} is called the *stress tensor*.

It is not a coincidence that T^{ij} is symmetric:

$$T^{ij} = T^{ji}.$$

It is related to conservation of angular momentum. But that is another story.

2.3   Incompressible flow

Start with the simplest case where ρ is a constant, independent of p. This is, of course, an approximation. It is valid when the velocity of the flow is small everywhere compared to the speed of sound. (See below for a derivation of the speed of sound.) Then the conservation of mass reduces to the statement that the divergence of the velocity field is zero.

$$\partial_i v^i = 0. \tag{2.2}$$

Such vector fields are said to be incompressible. The geometric meaning is that the volume of a small fluid element carried along the flow remains unchanged, even as its shape is distorted.

The conservation of momentum becomes

$$\frac{\partial v^i}{\partial t} + \partial_j \left[v^i v^j \right] = -\partial_i \left[\frac{p}{\rho} \right] + \frac{f_i}{\rho}.$$

Expanding the second term out

$$\partial_j \left[v^i v^j \right] = v^i \partial_j v^j + v^j \partial_j v^i.$$

The first term is zero by incompressibility.

$$\frac{\partial v^i}{\partial t} + v^j \partial_j v^i = -\partial_i \left[\frac{p}{\rho} \right] + \frac{f_i}{\rho}. \tag{2.3}$$

There are four equations (eqns 2.2, and 2.3) for four unknowns (the components of velocity and $\frac{p}{\rho}$). It is this particular case of incompressible flow that is often called Euler's equation, although Euler's original ideas were more general.

2.4   Conservation of energy

Using the identity from vector calculus

$$\frac{1}{2}\nabla v^2 = v \cdot \nabla v + v \times (\nabla \times v)$$

we get

$$\frac{\partial v}{\partial t} + \omega \times v + \nabla \left[\frac{p}{\rho} + \frac{1}{2}v^2 \right] = \frac{f}{\rho},$$

where $\omega = \nabla \times v$ is the vorticity. If the external force f is conservative, $f = -\nabla U$ (e.g., gravity) this becomes

$$\frac{\partial v}{\partial t} + \omega \times v + \nabla \left[\frac{p}{\rho} + \frac{1}{2}v^2 + \frac{U}{\rho} \right] = 0.$$

Taking a dot product and noting that $v \cdot \frac{\partial v}{\partial t} = \frac{\partial}{\partial t}\left[\frac{1}{2}v^2 \right]$ and $v \cdot [\omega \times v] = 0$, we get

$$\frac{\partial}{\partial t}\left[\frac{1}{2}v^2 \right] + v \cdot \nabla \left[\frac{1}{2}v^2 + \frac{p}{\rho} + \frac{U}{\rho} \right] = 0.$$

Using incompressibility $\nabla \cdot v = 0$, we can also write the above equation as

$$\frac{\partial}{\partial t}\left[\frac{1}{2}v^2 \right] + \nabla \cdot \left[\left\{ \frac{1}{2}v^2 + \frac{p}{\rho} + \frac{U}{\rho} \right\} v \right] = 0.$$

Integrating over some volume and applying Green's theorem we get

$$\frac{\partial}{\partial t}\int_V \frac{1}{2}v^2 \, dx + \int_{\partial V} \left\{ \frac{1}{2}v^2 + \frac{p}{\rho} + \frac{U}{\rho} \right\} v \cdot dS = 0.$$

If V is the whole domain of the fluid, the normal component of the velocity at the boundary must be zero. (The fluid cannot cross the boundary.) Then we get

$$\frac{\partial}{\partial t}H = 0, \quad H = \frac{1}{2}\int v^2 \, dx.$$

The total kinetic energy of an incompressible fluid is conserved. But we must remember that the ideal fluid is an approximation which is never quite fully realized: dissipation is present near the boundary even in the limit of small viscosity.

2.5   Helmholtz equation

If the fluid is incompressible, there is a one–one relation between ω and v: the equations

$$\nabla \times v = \omega, \quad \mathrm{div}v = 0$$

can be solved for v given ω and some boundary conditions. We can eliminate the pressure and potential energy from Euler's equation by taking a curl to get

$$\frac{\partial \omega}{\partial t} + \nabla \times [\omega \times v] = 0.$$

Recall the identity

$$\nabla \times [A \times B] = A(\nabla \cdot B) - B(\nabla \cdot A) + [B, A],$$

where

$$[B, A] \equiv (B \cdot \nabla)A - (A \cdot \nabla)B$$

is the commutator of vectors fields. Since $\nabla \cdot v = 0 = \nabla \cdot \omega$ we get the elegant equation of Helmholtz

$$\frac{\partial \omega}{\partial t} + [v, \omega] = 0.$$

It says that vorticity (thought of as a vector field) is preserved along the streamlines.

2.6   Steady flows

Steady (time-independent) flows represent the equilibrium states of the fluid. The molecules that make up the fluid are not at rest. But as one moves, another takes its place so that the overall state of the fluid is unchanged in time. From the Helmholtz equation we see that the spatial dependence is determined by

$$[v, \omega] = 0, \quad \nabla \cdot v = 0.$$

Even steady flows can be quite complicated, because of the nonlinearity of this system of equations.

If we require in addition that the vorticity is zero,

$$\nabla \times v = 0, \quad \nabla \cdot v = 0,$$

we get linear equations that are well understood. If the vorticity is zero ("irrotational flow") there is a scalar function ("velocity potential") such that

$$v = \nabla \phi.$$

Incompressibility then gives

$$\nabla^2 \phi = 0.$$

This Laplace equation, which most physicists associate with electrostatics these days, was first discovered in this context.

In two-dimensional flow, a more popular approach is to solve the incompressibility condition first:

$$\partial_1 v^1 + \partial_2 v^2 = 0$$

implies that there is a function A ("stream function") such that

$$v_1 = \partial_2 A, \quad v_2 = -\partial_1 A.$$

The condition for zero vorticity

$$\partial_1 v_2 - \partial_2 v_1 = 0$$

again becomes the Laplace equation for A

$$\partial_1^2 A + \partial_2^2 A = 0.$$

Solving the Laplace equation under different boundary conditions gave many of the early insights into fluid flow. We will see some examples later.

2.7   Equations of state

If the density is not a constant we need additional knowledge about the fluid to solve Euler's equation. The equation of continuity and Euler together give us four equations. But we have five unknowns: ρ, p, v^1, v^2, v^3. The last piece of information is the equation of state of the fluid, that is, a relation between ρ and p. This relation depends on the nature of the fluid, ultimately on its physical composition. There are three situations of special interest: constant density (which we looked at before), constant temperature, and constant entropy. Each is an approximation, valid in a different set of circumstances.

2.7.1   Constant temperature

The simplest equation of state is an ideal gas for which p and ρ are proportional:

$$p = \left[\frac{k_B T}{m} \right] \rho.$$

where k_B is Boltzmann's constant, m is the mass of one molecule of the gas, and T is the temperature. More generally, we can use the thermodynamic relation

$$\frac{dp}{\rho} - s dT = d\Phi,$$

where Φ is the Gibbs free energy and s is the entropy, per unit mass. At constant temperature

$$\frac{dp}{\rho} = d\Phi$$

so that

$$\frac{\partial \left[\rho v^i \right]}{\partial t} + \partial_j \left[\rho v^i v^j \right] = -\rho \partial_i \Phi + f_i.$$

To complete the story we still need a relation between Φ and ρ. Simplified models (called *polytropes*) are often good approximations:

$$\Phi = \alpha \rho^\gamma, \quad p = \gamma \alpha \rho^\gamma$$

for some parameters α, γ which may depend on temperature.

In general, the temperature of the fluid may not be constant: we get a mix of thermodynamics and fluid mechanics.

2.7.2   Constant entropy

A different approximation is to assume constant entropy rather than constant temperature. This allows for the heating of a fluid as it is compressed, without taking into account the loss of energy to dissipation (viscosity, radiative cooling, etc.). This is the right approximation when the change in volume is so rapid that the fluid does not have time to equalize its temperature, or increase its entropy. Recall that the enthalpy per unit mass is

$$w = \Phi + sT.$$

Thus,

$$dw = T ds + \frac{dp}{\rho}.$$

Since we are assuming that s is a constant, we get

$$dw = \frac{dp}{\rho}.$$

In this case Euler's equation becomes

$$\frac{\partial \left[\rho v^i\right]}{\partial t} + \partial_j \left[\rho v^i v^j\right] = -\rho \partial_i w.$$

To complete the circle we will need an equation of state that gives enthalpy $w(\rho, s)$ as a function of density and entropy. For polytropes

$$w = \alpha \rho^\gamma,$$

where α, γ depend on entropy.

All three cases above can be written as

$$\frac{\partial \left[\rho v^i\right]}{\partial t} + \partial_j \left[\rho v^i v^j\right] = -\rho \partial_i \psi + f_i,$$

where

$$\psi = \begin{cases} \frac{p}{\rho} & \text{constant density} \\ \Phi & \text{Gibbs free energy for constant temperature} \\ w & \text{enthalpy for constant entropy} \end{cases}$$

2.8 Transport form of Euler's equation

Expanding out the derivatives and using the conservation of mass,

$$\frac{\partial \left[\rho v^i\right]}{\partial t} + \partial_j \left[\rho v^i v^j\right] = \rho \frac{\partial v^i}{\partial t} - \partial_j \left[\rho v^j\right] v^i + \partial_j \left[\rho v^j\right] v^i + \rho v^j \partial_j v^i = \rho \left[\frac{\partial v^i}{\partial t} + v^j \partial_j v^i\right]$$

yielding

$$\frac{Dv^i}{dt} \equiv \frac{\partial v^i}{\partial t} + v^j \partial_j v^i = -\partial_i \psi + \frac{f_i}{\rho}, \tag{2.4}$$

$$\frac{\partial \rho}{\partial t} + \partial_j \left[\rho v^j \right] = 0.$$

....

Exercise 2.1 How would the equations change if the external force is due to Newtonian gravitational field?

Solution: The gravitational force on any fluid element is proportional to its mass. So, $f_i = -\rho \partial_i U$ where U is the gravitational potential. So we just have to add U to the thermodynamic potential: $\frac{\partial v^i}{\partial t} + v^j \partial_j v^i = -\partial_i [\psi + U]$.

....

....

Exercise 2.2 Derive a differential equation expressing the conservation of energy for an ideal compressible fluid.

Solution: In the absence of external forces $f_i = 0$, the total energy of an ideal fluid must be conserved. In the incompressible case, the energy of the fluid is entirely kinetic. In the compressible case, we must include the potential energy of compression. Proceeding by analogy to the incompressible case, take the scalar product wth velocity:

$$v_i \frac{\partial v^i}{\partial t} + v_i v^j \partial_j v^i = -v^i \partial_i \psi,$$

$$\frac{\partial}{\partial t} \left[\frac{1}{2} v_i v^i \right] + v^j \partial_j \left[\frac{1}{2} v_i v^i \right] = -v^i \partial_i \psi,$$

$$\frac{\partial}{\partial t} \left[\frac{1}{2} v_i v^i \right] + v^j \partial_j \left[\frac{1}{2} v_i v^i + \psi \right] = 0,$$

$$\rho \frac{\partial}{\partial t} \left[\frac{1}{2} v_i v^i \right] + \rho v^j \partial_j \left[\frac{1}{2} v_i v^i + \psi \right] = 0,$$

$$\frac{\partial}{\partial t} \left[\frac{1}{2} \rho v_i v^i \right] - \frac{\partial \rho}{\partial t} \frac{1}{2} v_i v^i + \rho v^j \partial_j \left[\frac{1}{2} v_i v^i + \psi \right] = 0.$$

Use the conservation of mass in the second term:

$$\frac{\partial}{\partial t} \left[\frac{1}{2} \rho v_i v^i \right] + \partial_j \left[\rho v^j \right] \frac{1}{2} v_i v^i + \rho v^j \partial_j \left[\frac{1}{2} v_i v^i + \psi \right] = 0,$$

$$\frac{\partial}{\partial t} \left[\frac{1}{2} \rho v_i v^i \right] - \psi \partial_j \left[\rho v^j \right] + \partial_j \left[\rho v^j \right] \left[\frac{1}{2} v_i v^i + \psi \right] + \rho v^j \partial_j \left[\frac{1}{2} v_i v^i + \psi \right] = 0,$$

$$\frac{\partial}{\partial t}\left[\frac{1}{2}\rho v_i v^i\right] - \psi \partial_j\left[\rho v^j\right] + \partial_j\left[\rho v^j\left(\frac{1}{2}v_i v^i + \psi\right)\right] = 0,$$

and again in the second term:

$$\frac{\partial}{\partial t}\left[\frac{1}{2}\rho v_i v^i\right] + \psi\frac{\partial\rho}{\partial t} + \partial_j\left[\rho v^j\left(\frac{1}{2}v_i v^i + \psi\right)\right] = 0.$$

The equation of state gives $\psi(\rho)$ as a function of density. Let Ψ be defined (up to an additive constant) by $\frac{d\Psi}{d\rho} = \psi$. Using

$$\frac{\partial\Psi\left(\rho(x,t)\right)}{\partial t} = \psi\left(\rho(x,t)\right)\frac{\partial\rho\left(\rho(x,t)\right)}{\partial t}$$

we get the conservation law

$$\frac{\partial}{\partial t}\left[\frac{1}{2}\rho v_i v^i + \Psi\right] + \partial_j\left[\rho v^j\left(\frac{1}{2}v_i v^i + \psi\right)\right] = 0.$$

Thus the total energy

$$H = \int\left[\frac{1}{2}\rho v_i v^i + \Psi\left(\rho\right)\right]dx$$

is conserved. We see that Ψ is the potential energy of compression. Again, we stress that this ideal case of conserved energy is never fully realized in nature: dissipation can never be neglected near the boundary.

....

....

Exercise 2.3 Derive the Helmholtz form of Euler's equation in the compressible case. That is, eliminate the gradient in eqn (2.4) by taking the curl and get an equation for vorticity (when $f_i = 0$).

Solution: As preparation, derive an identity of vector calculus.

$$\frac{1}{2}\partial_i\left[v_j v_j\right] = v_j \partial_i v_j$$

$$= v_j\left[\partial_i v_j - \partial_j v_i\right] + v_j \partial_j v_i$$

$$= \omega_{ij} v_j + v_j \partial_j v_i,$$

where ω_{ij} is the vorticity. So we have

$$v_j \partial_j v_i = -\omega_{ij} v_j + \frac{1}{2}\partial_j v^2$$

and hence

$$\frac{\partial v^i}{\partial t} - \omega_{ij} v_j + \partial_j\left[\frac{1}{2}v^2 + \psi\right] = \frac{f_i}{\rho}. \tag{2.5}$$

We can regard vorticity as a vector

$$\omega_1 = \partial_2 v_3 - \partial_3 v_2, \quad \omega = \text{curl} v$$

instead of an anti-symmetric tensor. Then

$$\frac{\partial v}{\partial t} - v \times \omega + \nabla \left[\frac{1}{2} v^2 + \psi \right] = \frac{f}{\rho}. \tag{2.6}$$

Taking a curl again (if the external force per density has zero curl as well) we get again the Helmholtz version of Euler's equation:

$$\frac{\partial \omega}{\partial t} = \text{curl} \left[v \times \omega \right], \quad \omega = \text{curl} v. \tag{2.7}$$

But this cannot be rewritten in terms of commutators because $\nabla \cdot v$ may not be zero.

2.9   Linearization of Euler's equations: sound

The Euler and continuity equations are quite nonlinear. The first thing to do with a nonlinear equation is always to look at its linearization around the simplest solution.

In our case this means small disturbances in a fluid at rest. Let p_0, ρ_0 be the equilibrium pressure and density (assumed to be independent of space as well as time) and $\check{p}, \check{\rho}$ the small disturbances. To first order in $\check{p}, \check{\rho}, v^i$ the continuity and Euler equations become

$$\frac{\partial \check{\rho}}{\partial t} + \rho_0 \partial_i v^i = 0,$$

$$\rho_0 \frac{\partial v^i}{\partial t} = -\partial_i \check{p}.$$

Combining these we get

$$\frac{\partial^2 \check{\rho}}{\partial t^2} - \partial_i \partial_i \check{p} = 0.$$

Linearizing the equation of state

$$\check{p} = \kappa_0 \check{\rho},$$

where

$$\kappa_0 = \left[\frac{\partial p}{\partial \rho} \right]_{\rho_0}$$

is the reciprocal of compressibility. So we get a wave equation

$$\frac{\partial^2 \check{\rho}}{\partial t^2} - \kappa_0 \partial_i \partial_i \check{\rho} = 0.$$

This means that density perturbations propagate as a wave, with speed

$$c_s = \sqrt{\kappa_0}.$$

(For example, a function $f(x^1 - \sqrt{\kappa_0} t)$ is a solution, describing waves propagating along the x^1-axis).) This is sound, the infinitesimal manifestation of fluid flow.

Since the period of sound waves is small compared to the time it takes a fluid to achieve constant temperature, it is best to use the constant entropy equation of state in calculating the speed of sound. Given the enthalpy

$$c_s = \sqrt{\left[\rho \frac{\partial w}{\partial \rho} \right]_{\rho = \rho_0}}.$$

In the approximation of an incompressible fluid $\frac{\partial \rho}{\partial p} = 0$, which means that the speed of sound is infinite. The correct interpretation is that it is large compared to the velocity of the fluid everywhere.

Sound waves carry the information about changes in pressure to the rest of the fluid. In an incompressible flow, this happens so fast that the fluid anticipates high pressure regions and flows around them. If the fluid velocity is larger than the speed of sound, it is unable to do that. The fluid can slam into a region of high pressure, and be forced to slow down suddenly. This can appear as a discontinuity in v, ρ, p called a "shock." The differential equations of fluid flow break down near shocks and have to be supplemented by additional conditions derived from the more fundamental physics of molecular dynamics or thermodynamics.

2.10 Inviscid Burgers equation

It is useful to think of the opposite limit from incompressibility, where the speed of sound goes to zero. In this limit, the pressure, instead of density, is a constant. The conservation of mass and momentum completely determine the flow:

$$\frac{\partial \rho}{\partial t} + \partial_i \left[\rho v^i \right] = 0,$$

$$\frac{\partial \left[\rho v^j \right]}{\partial t} + \partial_i \left[\rho v^i v^j \right] = 0.$$

We can eliminate ρ from the latter to get an equation involving the velocity alone:

$$\frac{\partial v^j}{\partial t} + v^i \partial_i v^j = 0.$$

The density is then advected along the flow determined by the above equation. We will return to studying this equation (mostly about shocks).

....

Exercise 2.4 Derive the differential version of conservation of energy for the inviscid Burgers equation.

Solution: Using the by now familiar procedure,

$$v_i \frac{\partial v^i}{\partial t} + v_i v^j \partial_j v^i = 0,$$

$$\frac{\partial}{\partial t}\left[\frac{1}{2} v_i v^i\right] + v^j \partial_j \left[\frac{1}{2} v_i v^i\right] = 0,$$

$$\frac{\partial}{\partial t}\left[\frac{1}{2} v_i v^i\right] + v^j \partial_j \left[\frac{1}{2} v_i v^i\right] = 0,$$

$$\rho \frac{\partial}{\partial t}\left[\frac{1}{2} v_i v^i\right] + \rho v^j \partial_j \left[\frac{1}{2} v_i v^i\right] = 0,$$

$$\frac{\partial}{\partial t}\left[\frac{1}{2} \rho v_i v^i\right] - \frac{\partial \rho}{\partial t}\frac{1}{2} v_i v^i + \rho v^j \partial_j \left[\frac{1}{2} v_i v^i\right] = 0,$$

$$\frac{\partial}{\partial t}\left[\frac{1}{2} \rho v_i v^i\right] + \partial_j \left[\rho v^j\right] \frac{1}{2} v_i v^i + \rho v^j \partial_j \left[\frac{1}{2} v_i v^i\right] = 0,$$

yielding the conservation law

$$\frac{\partial}{\partial t}\left[\frac{1}{2} \rho v_i v^i\right] + \partial_j \left[\rho v^j \frac{1}{2} v_i v^i\right] = 0.$$

Thus the total kinetic energy

$$H = \frac{1}{2} \int \rho v_i v^i dx$$

is conserved: there is no potential energy of compression. It is amusing that this holds in the limit of zero sound speed as well as infinite sound speed (incompressible flow).

....

2.11   Scale invariance

If we count the dimensions, $v \sim \frac{L}{T}$, $\partial \sim L^{-1}$. Thus the two terms on the l.h.s. of the incompressible Euler equation (eqn 2.3) both have dimensions of $\frac{L}{T^2}$. Pressure has units of energy density, and ρ is mass density so $\frac{p}{\rho}$ has units of the square of velocity. Thus the term on the r.h.s. of eqn (2.3), $\partial_i \left[\frac{p}{\rho}\right]$ also has units of $\frac{L}{T^2}$. Thus, in the absence of external forces the incompressible Euler equation has scale invariance:

$$x \to \lambda x, \quad t \to \tau t,$$

$$v \to \frac{\lambda}{\tau} v, \quad \frac{p}{\rho} \to \frac{\lambda^2}{\tau^2} \frac{p}{\rho}.$$

In addition, the equations are invariant under translations in space and time

$$x \to x + a, \quad t \to t + t_0,$$

and under rotations

$$x \to Rx, \quad R^T R = 1.$$

Any differential equation that depends only on the metric of Euclidean space (the Euler equation in particular) will have rotation and translation invariance in space.

Euler's equations are really a version of Newton's second law. Recall that it remains valid in any inertial reference frame. That is, given a solution, we can add a constant velocity to it and get another solution; provided we also change the position coordinate to the one measured by a moving observer:

$$x \to x + ut, \quad v \to v + u.$$

These are called Galilean transformations.

But this is not all. There is another symmetry you might not suspect. If we make the inversion in time

$$t \to -\frac{1}{t}$$

and supplement it with

$$x \to \frac{x}{t}$$

we get a symmetry. This somewhat surprising symmetry can be combined with time translation and scaling to get invariance under fractional linear transformations (also called conformal or Mobius transformations) in time (O'Raifeartaigh and Sreedhar, 2001):

$$t \to \frac{at + b}{ct + d}, \quad x \to \frac{x}{ct + d}, \quad ad - bc = 1.$$

For compressible fluids, there is still scale and conformal invariance (O'Raifeartaigh and Sreedhar, 2001) provided that the equations of state are of the polytropic form $p \propto \rho^\gamma$. You just need to transform density appropriately. That is, so that $\frac{p}{\rho}$ has dimensions of $\frac{L^2}{T^2}$:

$$\rho \sim \left(\frac{L}{T}\right)^{\frac{2}{\gamma-1}}.$$

2.11.1   The role of boundary conditions

The boundary conditions may not be invariant under the above transformations; their meaning is that the solutions with one set of boundary conditions are mapped to another solution with scaled (and translated) boundaries. For example, the flow past a large body can be related to that past a smaller body of the same shape. This principle was useful in the design of ships, airplanes, etc. where scale models were used to test out concepts. (Nowadays, numerical methods have replaced traditional wind tunnel testing.) It is still useful in finding laboratory models of astrophysical phenomena.

We will see that adding viscosity changes the story. A dimensionless quantity (Reynolds number) can be constructed out of velocity, the size of the boundary, and kinematic viscosity . Flows of the same Reynolds number are related by scaling. Again, this is useful in understanding large-scale phenomena by relating them to laboratory experiments.

2.12   d'Alembert's paradox: limitations the ideal fluid model

We should not end a discussion of ideal fluid theory without addressing its obvious flaw. It predicts that a symmetric body (e.g., a cylinder) placed in fluid flowing past it has no net drag force acting on it, contrary to experiments. In his famous *Lectures on Physics*, (Feynman, 2002) mocked the ideal fluid as the theory of "dry water." A related fact is that the boundary conditions on Euler equations only require that the normal component of the fluid velocity be zero (so that the fluid does not penetrate the boundary). But it is obvious experimentally that all components of velocity (including tangential) are zero at the boundary.

The point is that there is always some loss of energy from the collective motion of the fluids (described by the velocity field) to the microscopic molecular motion (which might cause an increase in temperature of the fluid). Wherever the fluid velocity is not a constant, different layers of fluid will rub against each other. This frictional force is proportional to the gradient of velocity. The proportionality constant is a property of the fluid, viscosity. A more accurate description of the fluid is the Navier–Stokes equation, which has a term proportional to viscosity. We will derive it in the next chapter.

This friction is most important near the boundary, where the velocity gradients are inevitably large. There has to be a boundary layer with large gradients for velocity to achieve zero tangential velocity at the boundary. It turns out that the net loss of energy in the boundary layer remains non-zero *even in the limit of zero viscosity*: the boundary layer becomes thin, but the velocity gradients become large to compensate for this. This boundary layer theory of Prandtl is an important addendum to ideal fluid theory (see Section 7.1). There are some further subtleties (separation of boundary layers) as well.

A parameter having a residual effect in the limit as it goes to zero happens in other areas of physics as well. Prandtl's boundary layer theory is a good model for developing an understanding of such "anomalies."

3

The Navier–Stokes Equations

Fluid dynamics deals with the collective motion of a large number of molecules, averaging over the motion of individual molecules in some tiny region. An example of the latter is the random (thermal) motion of molecules, which takes place over a region about the size of the mean free path between collisions. The energy in the collective motion can be transferred to that of the thermal motion (heat), which will appear as dissipation to fluid mechanics. This is a form of friction, called viscosity. Different layers of fluids rub against each other, transferring some fluid energy to heat. If the fluid velocity is a constant, friction must vanish: on average there is no transfer of momentum between molecules in the different layers during collisions. So also for fluids in uniform rotation ($v = \Omega \times r$ for constant Ω). That the energy of overall translation and rotation of a fluid is conserved follows from the symmetry of the scattering of molecules under translation and rotation. It is possible to derive the loss of energy in a systematic expansion in derivatives of velocity, but we will take a more phenomenological approach.

3.1 Viscosity

One way to understand this is to think of the flow of momentum across a small surface. Conservation of momentum is the statement

$$\frac{\partial [\rho v_i]}{\partial t} + \partial^j T_{ij} = f_j,$$

where T_{ij} is the stress tensor and f_j is the force acting on a fluid element. T_{ij} is the ith component of momentum flowing across a small surface whose normal points in the jth direction. In an ideal fluid, it is equal to

$$T_{ij} = \rho v_i v_j + p \delta_{ij}.$$

The loss of momentum due to friction ought to be represented by an extra term in the stress tensor. We want a simple model that does not delve deep into the molecular structure of the material. So we make the assumption that the viscous stress is linear in

Fluid Mechanics: A Geometrical Point of View. S. G. Rajeev © S. G. Rajeev 2018.
Published in 2018 by Oxford University Press. DOI: 10.1093/oso/9780198805021.001.0001

the velocity. (This is similar to the way we represent dynamic friction in mechanics by a force proportional to velocity.)

Since it must vanish for constant velocity, the viscous stress must depend on the derivatives $\partial_i v_j$ of velocity. It must also vanish for a fluid that is rotating uniformly. The different layers do not rub against each other in this case.

....

Exercise 3.1 Show that for a fluid in constant rotation $v = \Omega \times r$, the derivative $\partial_i v_j$ is anti-symmetric.

....

Geometrically, we are saying that the viscous stress must vanish when the velocity field is a Killing vector. That is, when the symmetric part of the derivative

$$\partial_i v_j + \partial_j v_i$$

is zero. The simplest model is that the stress is proportional to this symmetric derivative of velocity.

It is convenient to split this matrix into a part proportional to the identity and a traceless part:

$$\partial_i v_j + \partial_j v_i - \frac{2}{3}\delta_{ij}\partial^k v_k.$$

The quantity $\partial^k v_k = \text{div } v$ represents the compression of a small volume of fluid. The loss of energy due to this compression is at a different rate than the loss due to a strain in the velocity field. So we have a simple model for viscous stress:

$$T_{ij} = \rho v_i v_j + \tau_{ij},$$
$$\tau_{ij} = -\eta\left[\partial_i v_j + \partial_j v_i - \frac{2}{3}\delta_{ij}\partial^k v_k\right] - \zeta\delta_{ij}\partial^k v_k.$$

What is usually called viscosity is the coefficient η. The other coefficient ζ is not important for incompressible flow, but is relevant to the dissipation of sound and to supersonic flow (sometimes called "second viscosity"). This simple picture is empirically valid for many fluids.

There are also cases where it fails. For example, if the molecules have a very asymmetrical shape, with the length in one direction comparable to the mean free path for collisions. Then, the equations we derive here do not describe the experiments well. They are called non-Newtonian fluids. (This terminology is confusing as it suggests that Newton's laws are violated. They are not. A better name would have been non-Stokes fluids, as only Stokes law of viscous force is violated. Physicists are not always good at naming things.) We will not pursue this direction for now.

We are now ready to modify Euler's equation by including viscosity:

$$\frac{\partial \left[\rho v_i\right]}{\partial t} + \partial^j \left[\rho v_i v_j\right] - \partial^j \left[\eta \left\{\partial_i v_j + \partial_j v_i - \frac{2}{3}\delta_{ij}\partial^k v_k\right\} + \zeta \delta_{ij}\partial^k v_k\right] = -\partial_i p + f_i.$$

Equivalently,

$$\frac{\partial \left[\rho v_i\right]}{\partial t} + \partial^j \left[\rho v_i v_j\right] = \eta \nabla^2 v_i + \zeta' \partial_i \mathrm{div} v - \partial_i p + f_i,$$

where $\zeta' = \zeta - \frac{2}{3}\eta$.

These are called Navier–Stokes equations. Again, we have the continuity equation

$$\frac{\partial \rho}{\partial t} + \partial^i [\rho v_i] = 0.$$

As in the case of ideal fluids, to complete the Navier–Stokes and continuity equations, we need an equation of state relating density to pressure. We just need to add the viscous stress to the equations of motion of the ideal fluid.

....

Exercise 3.2 Derive the analog of eqn (2.4) for a viscous fluid.

Solution: Proceeding by analogy,

$$\frac{\partial \left[\rho v_i\right]}{\partial t} + \partial^j \left[\rho v_i v_j + p\delta_{ij}\right] = \partial^j \left[\eta \left\{\partial_i v_j + \partial_j v_i - \frac{2}{3}\delta_{ij}\partial^k v_k\right\} + \zeta \delta_{ij}\partial^k v_k\right],$$

$$\frac{\partial \rho}{\partial t}v_i + \rho\frac{\partial v_i}{\partial t} + \partial^i [\rho v_j] v_i + \rho v_j \partial^j v_i + \partial_i p = \partial^j \left[\eta \left\{\partial_i v_j + \partial_j v_i - \frac{2}{3}\delta_{ij}\partial^k v_k\right\} + \zeta \delta_{ij}\partial^k v_k\right],$$

$$\rho\frac{\partial v_i}{\partial t} + \rho v_j \partial^j v_i + \partial_i p = \partial^j \left[\eta \left\{\partial_i v_j + \partial_j v_i - \frac{2}{3}\delta_{ij}\partial^k v_k\right\} + \zeta \delta_{ij}\partial^k v_k\right],$$

$$\frac{\partial v_i}{\partial t} + v_j \partial^j v_i + \frac{1}{\rho}\partial_i p = \frac{1}{\rho}\partial^j \left[\eta \left\{\partial_i v_j + \partial_j v_i - \frac{2}{3}\delta_{ij}\partial^k v_k\right\} + \zeta \delta_{ij}\partial^k v_k\right],$$

yielding

$$\frac{\partial v_i}{\partial t} + v_j \partial^j v_i + \partial_i \psi = \frac{1}{\rho}\partial^j \left[\eta \left\{\partial_i v_j + \partial_j v_i - \frac{2}{3}\delta_{ij}\partial^k v_k\right\} + \zeta \delta_{ij}\partial^k v_k\right].$$

....

3.2   Viscous incompressible flow

For flow at constant density it is a bit simpler since div $v = 0$. After dividing by ρ,

$$\frac{\partial v_i}{\partial t} + v^j \partial_j v_i = \nu \nabla^2 v_i - \frac{1}{\rho} \partial_i p + f_i. \tag{3.1}$$

The dimensions of η are $\frac{\text{mass}}{\text{length}\times\text{time}}$. In studying fluids at constant density, the ratio

$$\nu = \frac{\eta}{\rho}$$

is what matters. This *kinematic viscosity* has dimensions $\frac{\text{length}^2}{\text{time}}$: the same as the diffusion constant. That makes sense, as it describes the diffusion of momentum through the fluid. The original coefficient η is also called *dynamical viscosity* to contrast with ν.

It should not surprise you that glycerine is more viscous than water: $\nu \approx 6.8 \text{ cm}^2 \text{ s}^{-1}$ vs $0.01 \text{ cm}^2 \text{ s}^{-1}$. But did you expect that air ($\nu \approx 0.15 \text{ cm}^2 \text{ s}^{-1}$) is more viscous than water? Mostly because it has much lower density.

3.3   Dissipation of energy at constant density

Recall that all the energy in an incompressible fluid is kinetic energy. Viscosity causes this energy to be lost, so its derivative must be negative. Let us calculate that rate.

$$\frac{\partial}{\partial t}\left[\frac{1}{2}v^2\right] + v^j \partial_j \left[\frac{1}{2}v^2\right] = \nu v_i \nabla^2 v_i - \frac{1}{\rho} v^i \partial_i p$$

$$\implies \frac{\partial}{\partial t}\left[\frac{1}{2}v^2\right] + \partial_j \left[\frac{1}{2}v^2 v_j\right] = \nu v_i \nabla^2 v_i - \frac{1}{\rho} \partial_i \left[p v_i\right].$$

Now

$$\partial_j \partial_j v_i = \partial_j \left[\partial_j v_i + \partial_i v_j\right]$$

since the last term vanishes by incompressibility. Thus

$$v_i \partial_j \partial_j v_i = v_i \partial_j \left[\partial_j v_i + \partial_i v_j\right]$$
$$= \partial_j \left[v_i \left(\partial_j v_i + \partial_i v_j\right)\right] - \partial_j v_i \left(\partial_j v_i + \partial_i v_j\right).$$

Using the symmetry of the last factor

$$\partial_j v_i \left(\partial_j v_i + \partial_i v_j\right) = \frac{1}{2}\left(\partial_j v_i + \partial_i v_j\right)\left(\partial_j v_i + \partial_i v_j\right).$$

Thus we have the identity

$$v_i \partial_j \left[\partial_j v_i + \partial_i v_j\right] = \partial_j \left[v_i \left(\partial_j v_i + \partial_i v_j\right)\right] - \frac{1}{2} \left(\partial_j v_i + \partial_i v_j\right) \left(\partial_j v_i + \partial_i v_j\right). \tag{3.2}$$

Thus

$$v_i \nabla^2 v_i = \partial_j \left[v_i \left(\partial_j v_i + \partial_i v_j\right)\right] - \frac{1}{2} \left(\partial_j v_i + \partial_i v_j\right) \left(\partial_j v_i + \partial_i v_j\right)$$

so that

$$\frac{\partial}{\partial t}\left[\frac{1}{2}v^2\right] + \partial_j \left[\left\{\frac{1}{2}v^2 + \frac{p}{\rho}\right\} v_j - \nu v_i \left(\partial_j v_i + \partial_i v_j\right)\right] = -\nu \frac{1}{2}\left(\partial_j v_i + \partial_i v_j\right)\left(\partial_j v_i + \partial_i v_j\right).$$

If it were not for the term on the r.h.s. this would have been a conservation law. The second term above is a total divergence. So its integral over a domain can be rewritten as an integral over the boundary. But this boundary integral is zero because the velocity (not just its normal component) vanishes there. We get that the energy decreases monotonically:

$$\frac{d}{dt}\int \frac{1}{2}v^2 d^3x = -\frac{1}{2}\nu \int \left(\partial_j v_i + \partial_i v_j\right)\left(\partial_j v_i + \partial_i v_j\right) d^3x. \tag{3.3}$$

The rate of loss of energy is proportional to the symmetric part of the gradient of velocity squared. In particular, a fluid with constant velocity or in uniform rotation does not lose energy. (Recall that this is exactly the principle we used to get the formula for viscous force.)

Thus, as soon as a large gradient in velocity develops, viscosity will act to dampen it. Solutions of Navier–Stokes are more regular than those of Euler. Yet we do not have a mathematically rigorous proof of regularity yet. More detailed bounds on the derivatives of velocity are needed. It is one of the "millenial" problems posed at the beginning of this century: one of the most famous problems in all of mathematics.

Even when ν is small, this dissipation cannot be ignored near a boundary: the velocity gradients will be large. Thus there is no such thing as an ideal fluid: dissipation at the boundary (at least) must always be taken into account.

3.4   Dissipation of energy for compressible flows

Starting with the Navier-Stokes equation for the transport of velocity (where again $\zeta' = \zeta - \frac{2}{3}\eta$),

$$\frac{\partial v_i}{\partial t} + v_j \partial^j v_i + \partial_i \psi = \frac{1}{\rho}\partial^j \left[\eta \left\{\partial_i v_j + \partial_j v_i\right\} + \zeta' \delta_{ij}\partial^k v_k\right],$$

we proceed as before to derive a scalar equation by taking the dot product:

$$v^i \frac{\partial v_i}{\partial t} + v^i v_j \partial^j v_i + v^i \partial_i \psi = \frac{1}{\rho} v^i \partial^j \left[\eta \left\{ \partial_i v_j + \partial_j v_i \right\} + \zeta' \delta_{ij} \partial^k v_k \right],$$

$$\rho \frac{\partial}{\partial t} \left(\frac{1}{2} v^i v_i \right) + \rho v^j \partial_j \left[\frac{1}{2} v^i v_i + \psi \right] = v^i \partial^j \left[\eta \left\{ \partial_i v_j + \partial_j v_i \right\} + \zeta' \delta_{ij} \partial^k v_k \right].$$

As for the ideal fluid the l.h.s. becomes:

$$\rho \frac{\partial}{\partial t} \left(\frac{1}{2} v^i v_i \right) + \rho v^j \partial_j \left[\frac{1}{2} v^i v_i + \psi \right]$$

$$= \frac{\partial}{\partial t} \left[\frac{1}{2} \rho v_i v^i + \Psi \right] + \partial_j \left[\rho v^j \left(\frac{1}{2} v_i v^i + \psi \right) \right],$$

where Ψ and $\psi = \frac{d\Psi}{d\rho}$ are (as before) given as function of ρ by the equation of state. We have the identities

$$v^i \partial^j \left[\eta \left\{ \partial_i v_j + \partial_j v_i \right\} \right] = \partial^j \left[\eta v^i \left\{ \partial_i v_j + \partial_j v_i \right\} \right] - \left[\partial^i v^j \right] \eta \left\{ \partial_i v_j + \partial_j v_i \right\}$$

$$= \partial^j \left[\eta v^i \left\{ \partial_i v_j + \partial_j v_i \right\} \right] - \frac{1}{2} \eta \left\{ \partial_i v_j + \partial_j v_i \right\} \left\{ \partial_i v_j + \partial_j v_i \right\}$$

and

$$v^j \partial_j \left[\zeta' \partial^k v_k \right] = \partial_j \left[\zeta' v^j \partial^k v_k \right] - \zeta' \left(\partial_j v^j \right)^2.$$

Taking the total derivatives of the l.h.s.,

$$\frac{\partial}{\partial t} \left[\frac{1}{2} \rho v_i v^i + \Psi \right] + \partial_j \left[\rho v^j \left(\frac{1}{2} v_i v^i + \psi \right) - \eta v_i \left(\partial_j v_i + \partial_i v_j \right) - \zeta' v^j \partial_k v^k \right]$$

$$= -\frac{1}{2} \eta \left(\partial_j v_i + \partial_i v_j \right) \left(\partial_j v_i + \partial_i v_j \right) - \zeta' \left(\operatorname{div} v \right)^2.$$

Integrating this we get the dissipation of energy

$$\frac{d}{dt} \int \left[\frac{1}{2} \rho v_i v^i + \Psi \right] dx = -\frac{1}{2} \eta \int \left(\partial_j v_i + \partial_i v_j \right) \left(\partial_j v_i + \partial_i v_j \right) dx - \zeta' \int \left(\operatorname{div} v \right)^2 dx.$$

3.5 Scale invariance: Reynolds number

The incompressible Euler equations are invariant scale transformations in space and time separately. Viscosity changes this story.

Let us consider incompressible flow first. Kinematic viscosity $v = \frac{\eta}{\rho}$ has dimensions of $\frac{L^2}{T}$; it describes the diffusion of momentum through the fluid. To keep v unchanged, and get invariance for the Navier–Stokes (NS) equations, we must link the scale changes of space and time:

$$x \to \tau^2 x, \quad t \to \tau t.$$

It is still possible to relate flows past bodies of different sizes, related by scaling. For example, consider flow past a body of size l (e.g., a sphere of radius l) which tends at infinity to a constant velocity v. The quantity

$$\mathcal{R} = \frac{vL}{v}$$

is dimensionless. Large Reynolds number corresponds to nearly ideal flow; for small Reynolds number dissipation dominates.

We can look at scale invariance another way. If we define dimensionless variables

$$v^i = V\tilde{v}^i, \quad x^i = L\tilde{x}^i, \quad t = \frac{L^2}{v}\tilde{t},$$

$$\frac{p}{\rho} = \frac{Lv}{L}\tilde{\psi},$$

the incompressible NS equations become

$$\frac{Vv}{L^2}\frac{\partial \tilde{v}_i}{\partial \tilde{t}} + \frac{V^2}{L}\tilde{v}^j\tilde{\partial}_j\tilde{v}_i = v\frac{V}{L^2}\tilde{\nabla}^2\tilde{v}_i - \frac{Vv}{L^2}\tilde{\partial}_i\tilde{\psi}.$$

Canceling factors we get the dimensionless form of the NS equations:

$$\frac{\partial \tilde{v}_i}{\partial \tilde{t}} + \mathcal{R}\,\tilde{v}^j\tilde{\partial}_j\tilde{v}_i = \tilde{\nabla}^2\tilde{v}_i - \tilde{\partial}_i\tilde{\psi}, \quad \operatorname{div}\tilde{v} = 0.$$

This shows that $\mathcal{R}$ measures the importance of the nonlinear or transport term relative to dissipation and pressure.

At large Reynolds numbers, many flows are no longer regular and instead appears to be random, varying rapidly in space and time. This transition from laminar to turbulent flow is one of the most puzzling phenomena in all of physics.

3.6   Navier–Stokes in curvilinear coordinates

We wrote the Navier–Stokes equations in Cartesian coordinates originally. This is defensible as they are the simplest coordinate system in Euclidean geometry. But often we need to solve for fluid flow with boundaries that are best described in curvilinear coordinates. It is best to learn a bit of Riemannian geometry (Rajeev, 2013) and rewrite the equations in arbitrary coordinates.

We will need the idea of the metric tensor, its inverse (the contravariant metric tensor), and the affine Christoffel symbols.

$$g^{ij}g_{jk} = \delta^i_k,$$

$$\Gamma^i_{jk} = \frac{1}{2}g^{id}\left[\partial_j g_{dk} + \partial_k g_{id} - \partial_d g_{jk}\right],$$

$$D_j v^i = \partial_j v^i + \Gamma^i_{jk}v^k,$$

$$\operatorname{div} v = D_i v^i = \frac{1}{\sqrt{\det g}}\partial_i\left[\sqrt{\det g}\,v^i\right].$$

Then the covariant forms of the Navier–Stokes equations are

$$\frac{\partial v^i}{\partial t} + v^j D_j v^i + g^{ij}\partial_j \psi + \frac{1}{\rho}D_j \tau^{ji} = 0,$$

$$\frac{\partial \rho}{\partial t} + D_i[\rho v^i] = 0,$$

with the viscous stress in n dimensions given by

$$\tau^{ij} = -\eta\left[D^i v^j + D^j v^i\right] - \left[\zeta - \frac{2}{n}\eta\right]\operatorname{div} v g^{ij}.$$

Here we think of the mass per unit volume as a scalar quantity, instead of a density in the geometric sense. That seems more convenient.

Example 3.1
***

In cylindrical polar coordinates

$$g_{ij}dx^i dx^j = dr^2 + r^2 d\phi^2 + dz^2,$$

and in spherical polar coordinates,

$$g_{ij}dx^i dx^j = dr^2 + r^2 d\theta^2 + r^2 \sin^2\theta d\phi^2.$$

3.7   Diffusion and advection

The Navier–Stokes (NS) equations are a nonlinear system of partial differential equations for velocity. It is useful to study a simpler, linear equation for a scalar field as a toy model. It is of independent interest anyway, as it describes how a scalar quantity (temperature, number density of a smoke particles, etc.) is advected by the velocity field.

In addition, we add a term analogous to viscosity, proportional to the second derivative of the scalar:

$$\frac{\partial \phi}{\partial t} + v \cdot \nabla \phi = \sigma \nabla^2 \phi.$$

σ has the meaning of heat conductivity if ϕ is temperature; or the diffusion constant if it is the concentration of some impurity. The second term describes the effect of fluid velocity and the r.h.s. the effect of random collisions with other molecules, much like smoke is carried along by wind, while it also spreads by diffusion through random collisions with the molecules of air. Note that σ has dimensions of $\frac{L^2}{T}$, just like kinematic viscosity v. In the NS equation, the analog of the second term $v \cdot \nabla v$ is nonlinear: velocity is advected along itself. Instead we imagine that a velocity field is given to us and we are asked to determine how ϕ evolves given v.

3.8   The diffusion kernel

The case $v = 0$ should be considered first. Then

$$\frac{\partial \phi}{\partial t} = \sigma \nabla^2 \phi \tag{3.4}$$

is the famous diffusion equation. The problem is to solve for $\phi(t, x)$ at some later time t given its initial value $\phi_0(x)$ at time $t = 0$. Because the equation is linear, we can reduce this problem to solving the case where $\phi_0(x)$ is concentrated at one point. Other cases can be solved by linear superposition. More precisely, suppose $K_t(x, y)$ is the solution to the initial value problem

$$\frac{\partial K_t(x, y)}{\partial t} = \sigma \nabla_x^2 K_t(x, y), \quad \lim_{t \to 0} K_t(x, y) = \delta(x, y). \tag{3.5}$$

Then the solution for the diffusion equation (eqn 3.5) with initial condition $\phi_0(x)$ is simply

$$\phi(t, x) = \int K_t(x, y) u_0(y) dy.$$

You can verify that the the equation is satisfied because $\phi(t, x)$ and $K_t(x, y)$ both satisfy the diffusion equation in the variable x; the initial condition follows from the defining property of the δ-function

$$\int \delta(x,y)u_0(y)dy = u_0(x).$$

The solution to eqn (3.5) can be found by Fourier analysis. But let us take a different approach, which will generalize to the case with advection.

First of all note that

$$\int K_t(x,y)dx = 1.$$

For, if we integrate the equation we get

$$\frac{\partial}{\partial t}\int K_t(x,y)dx = \int \nabla_x^2 K(x,y)dx.$$

But the r.h.s. is a total derivative and can be turned into a surface integral at infinity. The solution of interest will vanish at infinity, so this surface integral is zero. So the total "mass" of $K_t(x,y)$ is a conserved quantity. Its initial value is 1 from the defining property of the δ-function.

$$\lim_{t\to 0}\int K_t(x,y)dx = \int \delta(x,y)dx = 1 \implies \int K_t(x,y)dx = 1.$$

Thus K_t has dimensions of inverse volume, L^{-3}. Since σ has dimensions of $\frac{L^2}{T}$, we see that $[\sigma t]^{\frac{3}{2}}$ has dimensions of volume. Thus $[\sigma t]^{\frac{3}{2}} K_t(x,y)$ is a dimensionless quantity. Let us make the ansatz

$$K_t(x,y) = \frac{e^{-S_t(x,y)}}{[\sigma t]^{\frac{3}{2}}}$$

for some dimensionless quantity S_t. Differentiating each factor we get the identities

$$\frac{\partial K_t(x,y)}{\partial t} = \left\{-\frac{3}{2t} - \frac{\partial S_t(x,y)}{\partial t}\right\} K_t(x,y),$$

$$\nabla_x K_t(x,y) = \{-\nabla_x S_t(x,y)\} K_t(x,y).$$

Differentiating the latter equation

$$\nabla_x^2 K_t(x,y) = \left\{-\nabla_x^2 S_t(x,y)\right\} K_t(x,y) + [\nabla_x S_t(x,y)]^2 K_t(x,y).$$

Putting this into the diffusion equation and canceling out an overall factor of K_t,

$$-\frac{3}{2t} - \frac{\partial S_t(x,y)}{\partial t} = \sigma\,[\nabla_x S_t(x,y)]^2 - \sigma\nabla_x^2 S_t(x,y)$$

or

$$\frac{\partial S_t(x,y)}{\partial t} + \sigma\,[\nabla_x S_t(x,y)]^2 = \sigma \nabla_x^2 S_t(x,y) - \frac{3}{2t}.$$

The l.h.s. should be familiar from the Hamilton–Jacobi equation for a free particle in classical mechanics (apart from the constant σ).

Next, note that $K_t(x,y)$ can only depend on x, y through their difference $x - y$: the equation as well as its initial condition is unchanged if we translate x and y by the same constant vector a. Actually it can only depend only on the distance $|x - y|$ between the points as the equation is also invariant under a rotation of x and y. Thus $K_t(x,y)$ and hence $S_t(x,y)$ can only depend on $|x-y|$. But we also know that $S_t(x,y)$ is dimensionless, because it sits in the exponential. The only dimensionless quantity we can construct out of $|x - y|$, σ, and t is $\frac{|x-y|^2}{\sigma t}$. Since $\frac{|x-y|^2}{2t}$ is the solution to the Hamilton–Jacobi equation for a free particle, we are led to try the ansatz

$$S_t(x,y) = b + \frac{|x - y|^2}{c\sigma t}$$

for some constants b, c. We can put this into the above equation and use

$$\nabla_x |x - y|^2 = 2(x - y), \quad \nabla_x^2 |x - y|^2 = 2\nabla_x \cdot (x - y) = 2 \times 3$$

to get

$$-\frac{|x - y|^2}{c\sigma t^2} + \sigma\left[\frac{2(x - y)}{c\sigma t}\right]^2 = 2 \times 3\sigma \frac{1}{c\sigma t} - \frac{3}{2t},$$

$$\frac{|x - y|^2}{\sigma t^2}\left[-\frac{1}{c} + \frac{4}{c^2}\right] = \frac{3}{2t}\left[\frac{4}{c} - 1\right].$$

Thus both sides vanish if $c = 4$. The additive constant b drops out of the differentiations. It is determined by the normalization condition.

Working back to the beginning we have

$$K_t(x,y) = e^{-b}\frac{e^{-\frac{(x-y)^2}{4\sigma t}}}{[\sigma t]^{\frac{3}{2}}}.$$

The normalization condition $\int K_t(x,y)dx = 1$ now fixes the overall constant to give

$$K_t(x,y) = \frac{e^{-\frac{(x-y)^2}{4\sigma t}}}{[4\pi\sigma t]^{\frac{3}{2}}}.$$

Exercise 3.3 Show that $\int_{\mathbb{R}} e^{-u^2} du = \sqrt{\pi}$ for a single real variable. And that

$$\int \frac{e^{-\frac{(x-y)^2}{4\sigma t}}}{[4\sigma t]^{\frac{3}{2}}} dx = \int_{\mathbb{R}^3} e^{-u^2} du \text{ for a vector } u, \text{ by making the substitution } u = \frac{x-y}{\sqrt{4\sigma t}}.$$

Hence that $\int \frac{e^{-\frac{(x-y)^2}{4\sigma t}}}{[4\sigma t]^{\frac{3}{2}}} dx = \pi^{\frac{3}{2}}.$

This Gaussian distribution is so common in statistics that it is also called the normal distribution. Note that for small σ or t it is sharply peaked around the region $x \approx y$; in the limit $t \to 0$ it approaches the δ-function which has an infinite peak located at one point $x = y$. As t grows, $K_t(x, y)$ spreads out more evenly and as $t \to \infty$ it goes to zero everywhere. Thus if you place a drop of ink in a tub of water, at y at time $t = 0$, after a short time it will spread out over a region of size $\sim \sqrt{\sigma t}$. If you wait long enough it will spread out evenly over the whole of the tub. The ink density is now so small it is invisible.

3.9   Growth of entropy in diffusion

Diffusion is not time-reversible. Once the ink spreads, there is no way to put it back in the bottle. We can quantify this irreversibility by an inequality when the quantity ϕ is positive. If it describes a number density, positivity makes sense physically:

$$\frac{\partial \phi}{\partial t} = \sigma \nabla^2 \phi, \quad \phi(x, t) > 0.$$

Since $\nabla^2 \phi = \partial_i [\partial_i \phi]$ is a total divergence, $\int \phi dx$ is conserved. When $\phi > 0$ we can define a variant of this quantity, a kind of *entropy*:

$$S = -\int \phi(x, t) \log \phi(x, t) dx.$$

Its time derivative is

$$\frac{dS}{dt} = -\int \frac{\partial \phi}{\partial t} [1 + \log \phi(x, t)] dx$$

$$= -\sigma \int \nabla^2 \phi [1 + \log \phi(x, t)] dx$$

$$= \sigma \int \nabla \phi \cdot \nabla [1 + \log \phi(x, t)] dx$$

$$= \sigma \int \nabla \phi \cdot \frac{\nabla \phi}{\phi} dx.$$

Thus, if $\phi(x, t) > 0$ the entropy always increases:

$$\frac{dS}{dt} = \sigma \int \frac{|\nabla \phi|^2}{\phi} dx \geq 0$$

until ϕ becomes a constant and the gradient vanishes. The more spread out the distribution the greater the entropy.

....

Exercise 3.4 Calculate the entropy for the gaussian solution we found in the last section. See how it grows with time.

Solution We have

$$S = \int \frac{e^{-\frac{(x-y)^2}{4\sigma t}}}{[4\pi\sigma t]^{\frac{3}{2}}} \left[\frac{(x-y)^2}{4\sigma t} + \frac{3}{2} \log(4\pi\sigma t) \right] dx.$$

By the normalization condition the second term is just $\frac{3}{2} \log(4\pi\sigma t)$. The first integral can be the substitution $u = \frac{x-y}{\sqrt{4\sigma t}}$:

$$\int \frac{e^{-\frac{(x-y)^2}{4\sigma t}}}{[4\pi\sigma t]^{\frac{3}{2}}} \left[\frac{(x-y)^2}{4\sigma t} \right] dx = \int \frac{e^{-[u_1^2 + u_2^2 + u_3^2]}}{\pi^{\frac{3}{2}}} \left[u_1^2 + u_2^2 + u_3^2 \right] du_1 du_2 du_3.$$

Each term can be calculated:

$$\int \frac{e^{-[u_1^2 + u_2^2 + u_3^2]}}{\pi^{\frac{3}{2}}} \left[u_1^2 \right] du_1 du_2 du_3 = \int \frac{e^{-u_1^2}}{\sqrt{\pi}} u_1^2 du_1 = \frac{1}{2}.$$

Thus

$$S = \frac{3}{2} \left[1 + \log(4\pi\sigma t) \right].$$

It increases from $-\infty$ to ∞ as t varies from 0 to ∞.

....

3.10   The advection–diffusion kernel

Now let us see how the story changes with advection. We still seek a kernel that can give the solution

$$\phi(x, t) = \int K_t(x, y) \phi_0(y) dy$$

in terms of the initial data $\phi_0(y)$. It must satisfy

$$\frac{\partial K_t(x,y)}{\partial t} + v \cdot \nabla_x K_t(x,y) = \sigma \nabla_x^2 K_t(x,y), \quad \lim_{t \to 0} K_t(x,y) = \delta(x,y).$$

The simplest case is a constant velocity v. It is like smoke carried along by a constant wind while also spreading by diffusion. In the reference frame moving along with the wind, there is only diffusion; we solved that case in the last sub-section. So we can get the answer by transforming from this co-moving frame to the original frame:

$$x \to x - vt,$$

$$K_t(x,y) = \frac{e^{-\frac{(x-y-vt)^2}{4\sigma t}}}{[4\pi\sigma t]^{\frac{3}{2}}}.$$

You can check that it solves the advection–diffusion equation for constant v.

We should not expect to be able to solve the general case explicitly: the kernel should depend on the function $v(x, t)$. But we can get some approximations valid for small σ (or equivalently t).

Make the same ansatz as before (except that we sneak in the normalization constant determined earlier):

$$K_t(x,y) = \frac{e^{-S_t(x,y)}}{[4\pi\sigma t]^{\frac{3}{2}}}.$$

The advection–diffusion equation now takes the form

$$\frac{\partial S_t(x,y)}{\partial t} + v(x,t) \cdot \nabla_x S_t(x,y) + \sigma \, [\nabla_x S_t(x,y)]^2 = \sigma \nabla_x^2 S_t(x,y) - \frac{3}{2t}.$$

An approximate solution (valid when v is slowly varying) can be obtained by generalizing the solution for constant v. Note that $\xi(t) = x - vt$ is the solution to the ordinary differential equation

$$\frac{d\xi(t)}{dt} = -v, \quad \xi(0) = x.$$

When v is not a constant but is slowly varying we can solve instead the characteristic equation

$$\frac{d\xi(t)}{dt} = -v(t, \xi(t)), \quad \xi(0) = x.$$

That is, we look back along the flow to see where a dust particle could have come from at an earlier time. Then

$$K_t(x, y) \approx \frac{e^{-\frac{[x-y-\xi(x,t)]^2}{4\sigma t}}}{[4\pi\sigma t]^{\frac{3}{2}}}$$

is an approximate solution.

....

Exercise 3.5 Is the entropy still monotonically increasing for the advection–diffusion equation? Assume that the velocity field is incompressible $\nabla \cdot v = 0$ and that $\phi(x, t) > 0$.

Solution We can use $v \cdot \nabla = \nabla \cdot [v\phi]$ because of incompressibility:

$$\frac{dS}{dt} = -\int \frac{\partial \phi}{\partial t} [1 + \log \phi(x, t)] \, dx = -\int \left\{ -v \cdot \nabla \phi + \sigma \nabla^2 \phi \right\} [1 + \log \phi(x, t)] \, dx$$

$$= -\int \left\{ -\nabla[v \cdot \phi] + \sigma \nabla^2 \phi \right\} [1 + \log \phi(x, t)] \, dx$$

$$= -\int \nabla \cdot \{ -[v\phi] + \sigma \nabla \phi \} [1 + \log \phi(x, t)] \, dx$$

$$= \int \{ -[v\phi] + \sigma \nabla \phi \} \cdot \frac{\nabla \phi}{\phi} \, dx$$

$$= \int -v \cdot \nabla \phi \, dx + \int \sigma \nabla \phi \cdot \frac{\nabla \phi}{\phi} \, dx.$$

The first term is zero by integration by parts and the second is positive as we say earlier. Then

$$\frac{dS}{dt} > 0$$

if $\phi(x, t) > 0$ and $\nabla \cdot v = 0$. The rate of growth is controlled by diffusion.

....

4

Ideal Fluid Flows

We look at a few cases to illustrate some of the methods and to get a general idea of the underlying physics. Some of the examples will be of interest to us later, when we dive deeper into the dynamics of fluids.

4.1 Statics

When a flow is stationary $\left(\frac{\partial v}{\partial t} = 0\right)$, irrotational (curl $v = 0$), and if the external force per unit density is a gradient (as for gravity),

$$\frac{f}{\rho} = -\nabla U,$$

eqn (2.6) reduces to

$$\nabla \left[\frac{1}{2} v^2 + \psi + U \right] = 0$$

or

$$\frac{1}{2} v^2 + U + \psi = \text{constant}.$$

This is essentially the conservation of energy per unit density. The first term corresponds to the kinetic energy, the second potential energy, and the third the internal energy of the fluid. For a fluid at constant density we can also write this as

$$\frac{1}{2} v^2 + U + \frac{p}{\rho} = \text{constant},$$

which reminds us that pressure is a form of potential energy per unit volume. This is one of the oldest results in fluid mechanics: *Bernoulli's principle*.

Fluid Mechanics: A Geometrical Point of View. S. G. Rajeev © S. G. Rajeev 2018.
Published in 2018 by Oxford University Press. DOI: 10.1093/oso/9780198805021.001.0001

Example 4.1 Water tank

Bernoulli's principle can be used to understand the rate at which water drains out of a tank. For water in a tank, if the density and temperature are constant we get the elementary relation $p = \rho g h$. If we let the fluid flow out through a tube of cross-sectional area a at the base of a tank of area A, the rate at which the height of the fluid decreases is related to the fluid velocity in the tube: $A\frac{dh}{dt} = av$. Bernoulli gives $v = \sqrt{2gh}$. So $h = \frac{1}{2}\left[\frac{a}{A}\right]^2 gt^2$.

Example 4.2 Venturi tube

Bernoulli explains why a fluid at constant density *loses* pressure as it passes through a constriction in a pipe. To satisfy conservation of mass, as the cross-sectional area of the pipe decreases, the velocity must increase so that the same amount of mass crosses it. For a level pipe (so that gravitational potential is constant) the pressure must then decrease. Sensing the change in pressure Δp and the cross-sectional areas we can measure the volume per unit time Q of fluid flowing in the pipe:

$$Q = v_1 A_1 = v_2 A_2, \quad \frac{1}{2}v_1^2 + \frac{p_1}{\rho} = \frac{1}{2}v_2^2 + \frac{p_2}{\rho}, \quad \Delta p = p_2 - p_1$$

$$\implies Q = \sqrt{\frac{2}{\rho\left[A_1^{-2} - A_1^{-2}\right]}}\sqrt{\Delta p}.$$

Example 4.3 Pitot tube

Another application is the Pitot tube. Expose a tube to the flow of air around a subsonic airplane, so that some of the fluid is taken in and brought to rest. The pressure increase can be sensed with a gauge. The increase in pressure can be used to measure the air speed: $v = \sqrt{2\frac{\Delta p}{\rho}}$.

At supersonic speeds we have to take into account compressibility effects. The basic principle is the same, as long as we know the equation of state. Since the pressure changes are rapid, constant entropy is a reasonable approximation (rather than constant temperature: the air will get heated as it passes through shock waves):

$$v = \sqrt{2\left[w(p_2) - w(p_1)\right]},$$

where p_2 is the pressure in the tube and p_1 that outside.

Example 4.4 Self-gravitating fluids

For large bodies (stars, planets, etc.) we cannot assume that the acceleration due to gravity is a constant. When the fluid is at rest, $v = 0$, we have the conditions for hydrostatic equilibrium $U + \psi(\rho) = \text{constant}$, along with the Poisson equation (equivalent to Newton's law of gravity) $\nabla^2 U = 4\pi G\rho$.

To complete the story we need an equation of state. A polytropic equation of state $\psi = \alpha\rho^\beta$. An isothermal model $w = \alpha \log \rho$, which is the limiting case as $\beta \to 0$, is often a good approximation. Eliminating U and ρ using the above equations we get

$$\nabla^2 \psi = -4\pi G \left[\frac{\psi}{\alpha}\right]^n.$$

The quantity $n = \frac{1}{\beta}$ is called the polytropic index.

Assuming spherical symmetry (rotation is a small effect in many cases), this becomes a nonlinear ordinary differential equation for ψ. The solution can be found, using the boundary condition that the density is finite at the center and has a maximum there. Chandrasekhar's classic text (Chandrasekhar, 1961) discusses this in depth. A more modern approach is that by (Choudhuri, 2010).

Let ψ_0, ρ_0 be the central values (i.e., at $r = 0$) of ψ and ρ. In terms of the dimensionless variables

$$\theta = \frac{\psi}{\psi_0}, \quad \xi = \frac{r}{r_0},$$

where the length r_0 is given by

$$\frac{1}{r_0^2} = \frac{4\pi G\rho_0}{\psi_0} = \frac{4\pi G}{\alpha}\rho_0^{1-\frac{1}{n}},$$

we get the Lane–Emden equation

$$\theta'' + \frac{2}{\xi}\theta' = -\theta^n, \quad \theta(0) = 1, \quad \theta'(0) = 0.$$

The equation has physical meaning only when $\theta(\xi) \geq 0$. So it is valid for the range $0 \leq \xi \leq \xi_1$ where ξ_1 is the smallest root of $\theta(\xi_1) = 0$. For $0 < n < 5$ (the cases of most interest), this radius ξ_1 is finite.

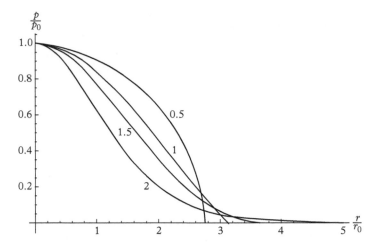

Figure 4.1 *The density profile of a self-gravitating fluid for different values of the polytropic index*

It is not difficult to solve this numerically (Figure 4.1), for various positive values of n. A trick is to impose the boundary condition at some small ξ (say 10^{-6}) instead of 0 to avoid the apparent singularity at the origin. You can convince yourself that $\theta(x) \sim 1 - \frac{1}{6}x^2$ for small ξ. So there is no true singularity at $\xi = 0$. In fact, $\theta = 1 - \frac{1}{6}\xi^2$ is the exact solution for the special case $n = 0$. Close to the origin, $\theta(\xi)$ tends to this for all n.

....

Exercise 4.1 Find a relation between the total mass and the central density. What is special about $n = 3$?

Solution The mass is $M = 4\pi \int_0^{r_1} \rho(r) r^2 dr$, where r_1 (the radius of the star) is the smallest root of the equation $\rho(r) = 0$. Passing to dimensionless variables

$$M = 4\pi r_0^3 \rho_0 \int_0^{\xi_1} \xi^2 \rho(\xi) d\xi,$$

putting in the formula for r_0:

$$M = 4\pi \left[\frac{4\pi G}{\alpha} \rho_0^{1-\frac{1}{n}} \right]^{-\frac{3}{2}} \rho_0 \int_0^{\xi_1} \xi^2 \rho(\xi) d\xi = \rho_0^{\frac{3-n}{2n}} 4\pi \left[\frac{\alpha}{4\pi G} \right]^{\frac{2}{3}} \int_0^{\xi_1} \xi^2 \rho(\xi) d\xi$$

When $n = 3$, the mass is independent of central density: it has a universal value, dependent only on the properties of the fluid (i.e., α) and not on the initial conditions. If the total mass is greater than this critical value, the star cannot exist in gravitational equilibrium and will collapse. This case arises for a relativistic degenerate electron gas and leads to the Chandrasekhar limit for the mass of a white dwarf star (Chandrasekhar, 2010; Choudhuri, 2010).

....

Exercise 4.2 Show that the Lane–Emden equation implies that $\int_0^\infty \xi^2 \theta^n(\xi) d\xi = -\xi_1^2 \theta'(\xi_1)$. Find its value for $n = 3$.

Solution We can write the equation as $\frac{d}{d\xi}\left[\xi^2 \frac{d\theta}{d\xi}\right] = -\xi^2 \theta^n$. Integrate both sides over $0 \le \xi \le \xi_1$ and use the boundary relations. Numerical solution gives $\xi_1 = 6.897$, and $-\xi_1^2 \theta'(\xi_1) = 2.018$ for $n = 3$.

Exercise 4.3 Find a relation between the mass and the radius of the star. Are more massive stars larger or smaller, according to this model with $1 \le n \le 3$?

4.2 Solutions of Laplace's equation

If $v = \text{grad } \phi$, the flow is irrotational: curl $v = 0$. If moreover, div $v = 0$ (constant density would imply this) we get

$$\nabla^2 \phi = 0.$$

This *Laplace's equation* appears all over physics. It was originally derived by Euler as this special case of his equations for a fluid. This was the only case people understood for many decades. We all recall fondly the many hours spent solving this equation for various boundary conditions. Only the time spent on a dentist's chair compares.

Many mathematical techniques, such as conformal transformations, were originally developed to solve these fluid problems. We will look at a couple of cases of interest in fluid mechanics.

4.2.1 A sphere moving through a fluid

A sphere of radius a moves through an ideal fluid at constant velocity u; the flow is steady, irrotational, and incompressible. So we must solve Laplace's equation. The boundary conditions are that the fluid is not allowed to cross the sphere. In other words, on the surface of the sphere, the normal components of the fluid velocity and the sphere velocity must be equal (Landau and Lifshitz, 1969).

Choose a coordinate system with the origin at the center of the sphere. This is an inertial coordinate system. Let r be the position vector in this system:

$$v = \nabla \phi,$$
$$\nabla^2 \phi = 0,$$
$$v \to 0, \quad |r| \to \infty,$$
$$\hat{r} \cdot v = \hat{r} \cdot u, \quad \text{if } |r| = a.$$

Recall that the solutions to the Laplace equation that vanish at infinity are $\frac{1}{r}$ and its derivatives. The only two vectors on which ϕ can depend are r, u. So we make the ansatz

$$\phi = \alpha u \cdot \operatorname{grad}\frac{1}{r} = -\alpha \frac{u \cdot r}{r^3},$$

$$v = -\alpha \frac{u}{r^3} + 3\alpha u \cdot r \frac{\hat{r}}{r^3}.$$

At the boundary

$$\hat{r} \cdot v = 2\alpha u \cdot \hat{r}\frac{1}{a^3} = \hat{r} \cdot u \implies \alpha = \frac{1}{2}a^3.$$

So the flow is

$$v_i = \frac{a^3}{2r^3}\left[-u_i + 3\hat{r}_i u \cdot \hat{r}\right].$$

Recognize that the velocity field outside the sphere is the same as the electric field of a dipole. You shouldn't take this solution seriously in the region just behind the sphere: there will usually be a turbulent wake; the assumptions that the velocity is irrotational and time-independent break down.

4.2.2 Flow past a wedge

Laminar flow (i.e., without turbulence) is very close to ideal: the effects of friction and viscosity are only important in a thin layer near the boundary. As preparation for studying this boundary layer, it is useful to solve the ideal fluid flow around some simple boundary shapes. We consider here a corner (Landau and Lifshitz, 1969).

This time choose cylindrical polar coordinates r, θ, z with axis at the corner. The planes that form the boundary are at $\theta = 0, \alpha$. The velocity is $v = \operatorname{grad}\phi$ and $\nabla^2\phi = 0$. The boundary condition that the flow be normal to the boundary is

$$\frac{\partial\phi}{\partial\theta} = 0, \quad \theta = 0, \alpha.$$

Recall that in these coordinates

$$\nabla^2\phi = \frac{\partial^2\phi}{\partial r^2} + \frac{1}{r}\frac{\partial\phi}{\partial r} + \frac{1}{r^2}\frac{\partial^2\phi}{\partial\theta^2} + \frac{\partial^2\phi}{\partial z^2}.$$

A solution is

$$\phi = Ar^n \cos n\theta$$

for some constants A, n. The velocity is best thought of in terms of radial and angular components:

$$v_r = Ar^n \cos n\theta, \quad v_\theta = -nAr^n \sin n\theta.$$

The b.c. gives

$$\sin n\alpha = 0 \implies n = \frac{\pi}{\alpha}.$$

A is determined by the flow at a large distance. The boundary must have an end; treating it as infinite along the radial direction is valid only near the corner.

4.3   Complex analytic methods

A special trick available in two dimensions is to combine the two components of a vector into a complex number.

The conditions for incompressibility and irrotationality,

$$\partial_1 v_1 + \partial_2 v_2 = 0, \quad \partial_1 v_2 - \partial_2 v_1 = 0,$$

can be combined into the single complex equation

$$[\partial_1 + i\partial_2]\left(v^1 - iv^2\right) = 0.$$

The real part is incompressibility and the imaginary part is irrotationality.

Defining the complex variables

$$z = x^1 + ix^2, \quad v = v^1 - iv^2,$$

this becomes the Cauchy–Riemann equation

$$\bar{\partial} v = 0.$$

That is, v is a complex analytic function of z. The velocity potential defined by $v_1 = \partial_2 A$, $v_2 = -\partial_1 A$ is also an analytic function

$$v(z) = i\frac{dA}{dz}.$$

The path $\left(\xi^1(t), \xi^2(t)\right)$ of a particle advected by the velocity field (the integral curve of the vector field) satisfies

$$\frac{d\bar{\xi}}{dt} = v(\xi(t)), \quad \xi(t) = \xi^1(t) + i\xi^2(t).$$

Note that the stream function is constant along the trajectory:

$$\frac{dA}{dt} = v^1 \partial_1 A + v^2 \partial_2 A = \partial_2 A \partial_1 A - \partial_1 A \partial_2 A = 0.$$

In ideal fluid theory, v is tangential at a boundary. (We will look at a more realistic boundary layer theory including viscosity later.) The Riemann mapping theorem says that by a complex analytic (conformal) transformation we can map any connected boundary to the circle. So, solving the problem for the circle and then applying a conformal map can solve any boundary value problem. We will work out a couple of examples to get a feel for this ancient technique.

4.4   Fluid with a stirrer

Imagine that we have introduced a thin rotating rod into an ideal fluid moving in a plane (Aref, 1984). Its axis is normal to the plane and the complex coordinate of its position is ζ. The rod is allowed to rotate around its axis, so the vorticity does not need to be zero at ζ: we can have a singularity there. The velocity must drop to zero far from the position of the rod: we wish to describe the situation where all the fluid motion is caused by the rod stirring the fluid. So, we seek an analytic function with a singularity at ζ and decreasing to zero at infinity. Moreover, the circulation $\int \oint v dz = \Gamma$, where Γ is a real number proportional to the rate of rotation of the rod.

All these conditions are satisfied by a simple pole:[1]

$$v = \frac{\Gamma}{2\pi i} \frac{1}{z - \zeta}.$$

The stream function is[2]

$$A = -\frac{\Gamma}{\pi} \mathrm{Re} \log [z - \zeta],$$

which may also be written as

$$A = -\frac{\Gamma}{\pi} \log |z - \zeta|.$$

---

[1] Note that in the physical situation, the velocity will never be infinite; this formula for v only holds outside the rod, which has a small finite radius.

[2] Recall that $\frac{\partial}{\partial z} \log z\bar{z} = \frac{1}{z}$ and that $\log z\bar{z} = 2\mathrm{Re} \log z$

The integral curves satisfy

$$\frac{d\bar{\xi}}{dt} = \frac{\Gamma}{2\pi i}\frac{1}{\xi(t) - \zeta}.$$

You can see that they are circles centered at ζ, by solving this equation. Or, just note that the stream function must be constant on them, so

$$|z - \zeta| = \text{constant},$$

which is the equation for a circle.

So far we are considering a fluid that fills the whole plane. But the solution is still valid if the fluid is contained within a circle centered at ζ: the velocity is tangential to the boundary.

4.4.1   Fluid in the upper half-plane with a stirrer

Suppose now the boundary is the real axis. The fluid fills the upper half-plane with a stirrer at ζ.

We want an analytic function with a simple pole at ζ and no other singularity in the upper half-plane. Moreover the function is real along the real z-axis (velocity is tangential to the boundary). These conditions are satisfied by a difference of simple poles (Figure 4.2):

Figure 4.2 *Flow around a stirrer in the half-plane*

$$v(z) = \frac{\Gamma}{2\pi i}\left[\frac{1}{z - \zeta} - \frac{1}{z - \bar{\zeta}}\right]$$

if ρ is purely imaginary. The stream function is

$$A(z) = -\frac{\Gamma}{\pi}\mathrm{Re}\log\frac{z - \zeta}{z - \bar{\zeta}}.$$

This is an example of the method of images: we pretend that the fluid extends over the entire complex plane, and place a stirrer rotating in the opposite direction at the point reflected across the boundary. The integral curves are still circles, but they are not centered at ζ anymore. Instead, we have (since A is a constant along trajectories)

$$\left|\frac{z - \zeta}{z - \bar{\zeta}}\right| = \lambda$$

for some constant λ determined by the initial position. This can be rewritten as

$$|z - \zeta|^2 = \lambda^2 |z - \bar{\zeta}|^2,$$

$$|z - z_c|^2 = r^2, \quad z_c = \frac{\zeta - \lambda\bar{\zeta}}{1 - \lambda^2}, \quad r = \sqrt{|z_c|^2 - |\zeta|^2}.$$

4.4.2   Stirrer inside a circle

The Cauchy–Riemann equation is invariant under conformal transformations, which replace z by a complex analytic function: $z \mapsto w(z)$. An example is a Mobius transformation (also called fractional linear transformation):

$$w = i\frac{z - a}{z + a},$$

where a is a positive number. Its inverse is also a Mobius transformation:

$$z = a\frac{w - i}{w + i}.$$

The upper half-plane $\mathrm{Im}\, w > 0$ is mapped to the disc $|z| < a$; the real axis in the w-plane goes to the circle of radius a the z-plane.

Using this trick we can get the solution for a stirred fluid inside a disc of radius a centered at the origin. Functions transform simpler than vector fields. So we apply the transformation to the stream function from the last section to get

$$A(z) = -\frac{\Gamma}{\pi}\mathrm{Re}\log\frac{i\frac{a+z}{a-z} - i\frac{a+\zeta}{a-\zeta}}{i\frac{a+z}{a-z} + i\frac{a+\bar{\zeta}}{a-\bar{\zeta}}},$$

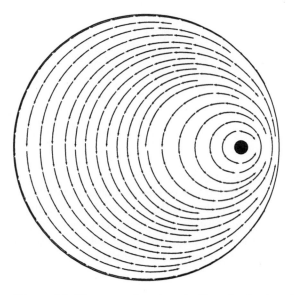

Figure 4.3 *Flow around a stirrer inside a circular boundary*

which simplifies to

$$A(z) = -\frac{\Gamma}{\pi}\operatorname{Re}\log\frac{z-\zeta}{z-\frac{a^2}{\bar\zeta}} + \text{constant.}$$

We don't need to calculate the additive constant, as it doesn't affect the flow. The corresponding velocity field is

$$v(z) = \frac{\Gamma}{2\pi i}\left\{\frac{1}{z-\zeta} - \frac{1}{z-\frac{a^2}{\bar\zeta}}\right\}.$$

The integral curves are again circles

$$\left|\frac{z-\zeta}{z-\frac{a^2}{\bar\zeta}}\right| = \lambda,$$

where λ is determined by the initial position of the trace particle (initial condition for the integral curve).

$$\lambda = \left| \frac{z_0 - \zeta}{z_0 - \frac{a^2}{\zeta}} \right|.$$

We can rewrite this as

$$|z - z_c| = r, \quad z_c = \frac{\zeta - \lambda^2 \frac{a^2}{\zeta}}{1 - \lambda^2}, \quad r = \frac{\lambda}{1 - \lambda^2} \left| \frac{a^2}{\zeta} - \zeta \right|.$$

The center z_c and radius r vary with the initial position of the tracer particle. Also, the tracer particle does not go around this circle at a constant angular velocity.

In Figure 4.3, the black dot is the position of the stirrer, and the arrows go around the circular path of tracer particles. Notice how the velocity field is tangential at the boundary. Of course, in reality, there will be a boundary layer in between this flow and the wall: the velocity would be zero at the wall not just tangential to it.

4.5   Flow past a cylinder

A fluid is moving with constant velocity u. Find how the fluid velocity changes when you place an infinite cylinder of radius a in it, with axis normal to u. Use the incompressible ideal fluid approximation.

The fluid will go around the cylinder, with zero vorticity. Translation along the axis of the cylinder is a symmetry, so it is enough to solve the problem in the plane normal to it. The boundary condition is that the fluid velocity should be tangential to the circle of radius a.

We can solve this by analogy to the sphere in the earlier section, except now we have a circle in a plane. r is now the cylindrical radial coordinate:

$$v = \nabla \phi,$$
$$\nabla^2 \phi = 0,$$
$$v \to u, \quad |r| \to \infty,$$
$$\hat{r} \cdot v = 0, \quad \text{if } |r| = a.$$

The analog of $\frac{1}{r}$, solving the Laplace equation in the plane, is $\log r$ and the velocity tends to u at infinity. So,

$$\phi = u \cdot r + \alpha u \cdot \nabla \log r = \alpha \frac{u \cdot r}{r^2},$$
$$v = u + \alpha \left[\frac{u}{r^2} - 2 \frac{u \cdot \hat{r}}{r^2} \hat{r} \right].$$

The boundary condition gives

$$u \cdot \hat{r} + \alpha \left[\frac{u \cdot \hat{r}}{a^2} - \frac{2u \cdot \hat{r}}{a^2} \right] = 0 \implies \alpha = a^2.$$

Thus (Figure 4.4),

$$v = u + \frac{a^2}{r^2} \left[u - 2\hat{r}\, u \cdot \hat{r} \right].$$

If u is along the x-axis we can write this explicitly as

$$v = u \begin{pmatrix} -\frac{a^2}{x^2+y^2} + \frac{2a^2 y^2}{(x^2+y^2)^2} + 1 \\ -\frac{2a^2 xy}{(x^2+y^2)^2} \end{pmatrix}.$$

It is interesting to see how the same problem is solved by the complex analysis method. We solve the condition for incompressibility first so that $v_1 = \partial_2 A, v_2 = -\partial_1 A$ or $v \equiv v_1 - iv_2 = [\partial_2 + i\partial_1] A = i [\partial_1 - i\partial_2] A$.

The stream function is a real solution to the Laplace equation; it is the real part of some complex analytic function. Before we place the cylinder, the velocity is a constant u in (say) the x-direction. Then $A_0 = uy = \mathrm{Re}[-iu(x+iy)] = \mathrm{Re}[-iuz]$. Placing the cylinder into the flow will add some harmonic function to it that must vanish at infinity (so that the flow returns to the original value at infinity). Moreover A must be constant

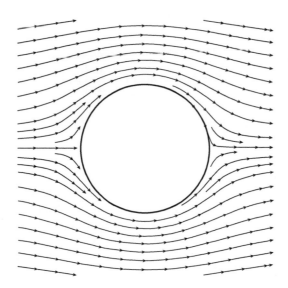

Figure 4.4 *Flow around a cylinder*

on the circle $z\bar{z} = a^2$ so that the fluid velocity is tangential to the circle (i.e., the circle must be a streamline.) These conditions are satisfied if

$$A = \text{Re}\left[-iu\left\{z - \frac{a^2}{\bar{z}}\right\}\right].$$

The second term is the "mirror image" of the first under the conformal transformation that preserves the circle. That is,

$$A = uy\left[1 - \frac{a^2}{x^2 + y^2}\right].$$

Differentiating and simplifying,

$$v = u\left[1 - \frac{a^2}{(x + iy)^2}\right].$$

We can now find the pressure using Bernoulli's theorem:

$$p(x, y) + \frac{1}{2}\rho|v(x, y)|^2 = p_0,$$

where p_0 is the constant pressure at infinity. That is,

$$p(x, y) = p_0 - \frac{1}{2}\rho u^2 \frac{(a^2 - 2ax + x^2 + y^2)(a^2 + 2ax + x^2 + y^2)}{(x^2 + y^2)^2}, \quad x^2 + y^2 \geq a^2.$$

On the circle it simplifies to

$$p = p_0 - 2\rho u^2\left[1 - \frac{x^2}{a^2}\right].$$

4.6   The d'Alembert paradox

The above pressure is symmetric under the reflection $x \to -x$. That is, the pressure on the front edge of the cylinder is the same as on the trailing edge. In particular, there is no drag on the cylinder, that is, no net force in the direction of the wind. This is clearly wrong physically and is an artifact of the ideal fluid approximation. In the real world, the flow will break up into an irregular pattern near the trailing edge: this "wake" will cause a low pressure region there, which causes a drag force on the cylinder. A more careful analysis shows that even in the limit of small viscosity, there is a small boundary layer where there are large velocity gradients. The fluid velocity at the boundary should be zero: even the tangential component must vanish. Moreover, the boundary layer separates from the cylinder, causing the wake.

4.7   Joukowski airfoil

The Cauchy–Riemann equations are unchanged under complex analytic (conformal) transformations $z \mapsto \zeta(z)$. The Riemann mapping theorem assures us that using such a map we can map the circle to any simple (i.e., without self-intersections) closed curve we want. Thus we can take the solution we found for flow past a cylinder and transform it to flow past a cylinder whose cross-section is any shape we want. This can be a first approximation to understanding the aerodynamics of a wing, if we model air as an ideal fluid and the wing is very long. Despite the fact that ideal fluid theory underestimates the drag, it gives a good understanding of lift (the component of force perpendicular to the asymptotic fluid velocity).

These simple mathematical models of wings were useful in the early days of aviation. Much more accurate numerical models go into the design of modern airplanes. Still, it is fun to revisit this ancient theory.

The Joukowski map is a clever transformation that maps circle in the ζ-plane to something that looks plausibly like an airfoil in the z-plane:

$$z = r\zeta + c + \frac{1}{r\zeta + c}.$$

The parameters $r > 0$ and the complex number c are parameters that can be chosen to get interesting shapes. It is convenient to re-express them as

$$c = -h + ib(1 + k), \quad r = (1 + k)\sqrt{1 + b^2},$$

and b, k are positive parameters less than one. It is possible to understand this transformation as a composition of conformal transformations. But it is simpler these days to just plot the curve $z\left(e^{i\theta}\right)$ (e.g., using ParametricPlot in Mathematica) and play around with different values of b, h or c, r. The case shown in Figure 4.5 is $b = 0.15$, $h = 0.075$, or $r = 1.09, c = -0.08 + 0.16i$.

To find the velocity distribution, we need the inverse transformation:

$$\zeta(z) = \frac{-2cr - r\sqrt{z^2 - 4} + rz}{2r^2}.$$

The velocity field around a unit circle is

$$v(\zeta) = u\left[1 - \frac{1}{\zeta^2}\right].$$

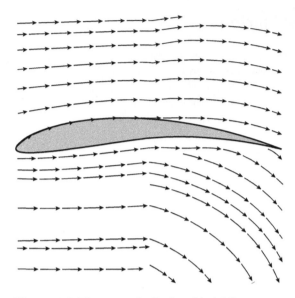

Figure 4.5 *Flow around a Joukowski airfoil*

In applying the transformation, we must remember that v transforms like a covariant vector field (or 1-form). That is,

$$v(\zeta)d\zeta = v\left(\zeta(z)\right)\zeta'(z)dz \equiv V(z)dz.$$

If we plot $V(x+iy) = V_1(x,y) - iV_2(x,y)$ we get the flow around the airfoil. The explicit calculations are best done by symbolic computation (e.g., Mathematica).

4.8   Surface waves

Let us look for a time-dependent solution next: waves on the surface of water. At equilibrium, the surface is a plane. Any disturbance will move as a wave with some constant speed. What is this speed? When the surface is displaced, it is restored by gravity. So the speed must involve the acceleration due to gravity g. To get a quantity with dimension of speed we also need some length l. Dimensionally, $c \sim \sqrt{gl}$. This l is the wavelength λ or the depth d, whichever is smaller. In most systems, the slowest waves are the most evident, as they cost the least energy to excite. We now do a more detailed analysis.

For irrotational incompressible flow, the vorticity form of Euler's equations becomes just

$$\frac{\partial}{\partial t}\partial_i\phi + \partial_i\left[\frac{1}{2}v^2 + \frac{p}{\rho} + U\right] = 0,$$

so that

$$\frac{\partial \phi}{\partial t} + \frac{1}{2}(\nabla \phi)^2 + \frac{p}{\rho} + U = F(t),$$

the function F of time alone being determined by the boundary conditions. The function ϕ must satisfy the Laplace equation for irrotational incompressible flow:

$$\nabla^2 \phi = 0,$$

so it is completely determined in the bulk by the boundary value. So the dynamical degrees of freedom are associated with the boundary.

In usual boundary value problems, we determine the value of some field in the interior knowing its value at a fixed boundary. Problems where the boundary values are themselves dynamical are especially difficult: only the simplest such problems can be treated analytically. Surface waves are simple to understand only if we make the approximation of ignoring the nonlinear terms in the Euler equations. This is valid as long as the amplitude of the wave is small compared to its wavelength. (Nonlinear phenomena like solitons will need deeper methods.)

So ignoring the nonlinear term we get the equations

$$\nabla^2 \phi = 0, \quad \frac{\partial \phi}{\partial t} + \frac{p}{\rho} + gz = F(t),$$

valid in the bulk of the fluid. We need boundary conditions to solve these equations.

In the absence of waves, the boundary is flat: $z = 0$ (say). We can choose the pressure on this plane to be zero by adding a constant to it. Then $F(t) = 0$. If there is a small displacement ζ of the surface,

$$g\zeta + \left[\frac{\partial \phi}{\partial t}\right]_{z=0} = 0.$$

At the boundary, the velocity of the fluid must be equal to the velocity of the boundary:

$$\left[\frac{\partial \phi}{\partial z}\right]_{z=0} = \frac{\partial \zeta}{\partial t}.$$

Thus we need to solve the Laplace equation in the bulk with the boundary condition

$$\left[g\frac{\partial \phi}{\partial z} + \frac{\partial^2 \phi}{\partial t^2}\right]_{z=0} = 0.$$

Suppose the depth of the water is d. This means that the vertical component of velocity must vanish at that second boundary:

$$\left[\frac{\partial \phi}{\partial z}\right]_{z=-d} = 0.$$

Let us make a separable ansatz for a wave moving in the x-direction with wavenumber k and frequency ω.

$$\phi = f(z)\cos[kx - \omega t].$$

The Laplace equation and its boundary condition at the bottom become

$$f'' - k^2 f = 0, \quad f'(-d) = 0 \implies f(z) = A\cosh k[z + d].$$

At the surface boundary

$$\left[gf' - \omega^2 f\right]_{z=0} = 0 \implies \omega = \sqrt{gk\tanh[kd]}.$$

This is an example of a dispersion relation. That is, how the frequency depends on the wavelength determines many of its properties. For example, the group velocity (the speed with which the center of a wave packet of finite width moves) is $c_g = \frac{d\omega}{dk}$. The phase velocity (the change of the phase of a plane wave) is $c = \frac{\omega}{k}$. They are the same only in the simplest case $\omega \propto k$.

In our case

$$c = \sqrt{gd}\sqrt{\frac{\tanh[kd]}{kd}}, \quad c_g = \frac{1}{2}\sqrt{gd}\frac{[kd]\text{sech}^2[kd] + \tanh[kd]}{\sqrt{[kd]\tanh[kd]}}.$$

For small k (long wavelength),

$$c_g \approx \sqrt{gd},$$

while for large k (small wavelength λ compared to d)

$$c_g \approx \frac{1}{2}\sqrt{\frac{g}{k}} = \frac{1}{2}\sqrt{\frac{g\lambda}{2\pi}}.$$

Longer waves (smaller k) move faster, as you can see in Figure 4.6. For wavelength of a meter (and much larger depths),

$$c_g \approx 0.6 \text{ m s}^{-1} \approx 2 \text{ km h}^{-1}.$$

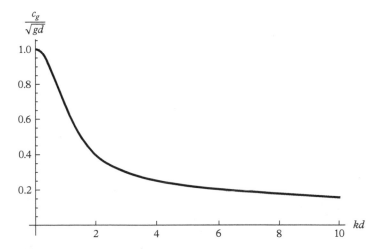

Figure 4.6 *Group velocity of surface waves*

There are ocean waves with wavelengths as large as $\lambda \sim 100$ km, much more than the depth of even the deepest trenches ($d \sim 10$ km). They move as fast as jet airplanes:

$$c_g \approx 200 \text{ m s}^{-1} \approx 720 \text{ km h}^{-1}.$$

The most spectacular of these waves are tsunamis. In the open ocean, they have an amplitude of only about 1 m, justifying our linear approximation. The extremely long wavelength means they carry a huge amount of energy, as a large body of water is being displaced. This energy comes usually from an earthquake, which causes a landslide extending over about a 100 km. As they approach the shore, the wave slows down and the amplitude grows: conservation of volume forces the water to rise to greater heights. The linear approximation is no longer valid. The nonlinear equations have soliton solutions that describe these waves.

The much more familiar, if less impressive, water waves are of small wavelength and even smaller amplitude. They move very slowly, the speed dropping to zero in the limit of small wavelength. If you pay attention while surfing or sailing you will notice that the longer waves move faster and catch up to the shorter waves ahead of them.

5

Viscous Flows

5.1 Pipe Poiseuille flow

Consider a viscous fluid flowing steadily along a long horizontal pipe whose cross-section is a circle of radius a. Choose cylindrical polar coordinates with the z-axis to point along the pipe.

Velocity should point along the axis of the pipe everywhere, and must vanish at the boundary. Also, in the limit of an infinitely very long pipe, it will be independent of z and ϕ. So we make the ansatz

$$v = \begin{pmatrix} 0 \\ 0 \\ v_z(r) \end{pmatrix}.$$

The pressure on the other hand will decrease along the pipe, $p(r, z)$, and be independent of the angle.

The Navier–Stokes (NS) equation reduces to

$$\frac{\partial p(r, z)}{\partial r} = 0,$$

$$\frac{\partial p(r, z)}{\partial z} - \nu \left[v_z''(r) + \frac{1}{r} v_z'(r) \right] = 0.$$

So the pressure depends only on z. The second equation separates into terms depending only on z and only on r. So each must be constant. So pressure is a linear function of z:

$$p(z) = -\frac{|\Delta p|}{L} z,$$

where L is the length of the pipe and $|\Delta p|$ the magnitude of the pressure difference from one end to the other.

Fluid Mechanics: A Geometrical Point of View. S. G. Rajeev © S. G. Rajeev 2018.
Published in 2018 by Oxford University Press. DOI: 10.1093/oso/9780198805021.001.0001

Also,

$$v_z''(r) + \frac{1}{r}v_z'(r) = -\frac{|\Delta p|}{\nu L}.$$

The solution which vanishes at $r = a$ is

$$v_z(r) = \frac{|\Delta p|}{4\nu L}\left[a^2 - r^2\right].$$

5.2 Circular Couette flow

Another classic example (Chandrasekhar, 1961) is a fluid trapped between two rotating circular cylinders with a long common axis. Let the radii be $a_1 > a_2$ and the angular velocities ω_1, ω_2.

Again, we choose the cylindrical polar coordinates. By the symmetries of the problem, we can look at first for a solution with

$$v = \begin{pmatrix} 0 \\ v_2(r) \\ 0 \end{pmatrix}.$$

That is, the velocity is purely angular v_2 and it depends only on r. It turns out that this simple solution (found by Stokes in 1840s) has *instabilties*, as shown by the experiments of Taylor in the 1920s. But this ansatz is still a good place to start understanding this problem.

The radial component of the NS equation reduces to

$$\frac{dp}{dr} = \rho r v_2^2(r).$$

The angular component gives

$$v_2''(r) + \frac{3}{r^2}v_2'(r) = 0 \implies v_2(r) = A + \frac{B}{r^2},$$

and the solution of the pressure equation is

$$p = \rho\left[\frac{1}{2}A^2 r^2 + 2AB\log r - \frac{1}{2r^2}B^2\right] + \text{constant}.$$

At the boundary the fluid must have the same angular velocity as the surface,

$$v_2(r) = \omega_{1,2}, \quad r = a_{1,2},$$

which determines the constants of integration,

$$A = \frac{a_2^2\omega_2 - a_1^2\omega_1}{a_2^2 - a_1^2}, \quad B = \frac{a_1^2 a_2^2 (\omega_1 - \omega_2)}{a_2^2 - a_1^2}.$$

5.3   Stokes flow

In the limit of large viscosity or small R we can ignore the nonlinear (transport) term to get

$$\frac{\partial v_i}{\partial t} = \nu \nabla^2 v_i - \frac{1}{\rho}\partial_i p + f_i, \quad \text{div } v = 0, \quad \text{Stokes equations.}$$

Being linear, they are much easier to solve. They describe the flow of thick fluids like molasses or lubricating oil, and even water, for very small time intervals, for which the nonlinear effects have not kicked in yet.

For steady flow at small Reynolds number and without external forces, the viscous force is balanced by the pressure gradient:

$$\eta \nabla^2 v = \nabla p, \quad \text{div } v = 0.$$

This implies $\nabla^2 p = 0$ and we might as well eliminate pressure to solve the fourth order equation

$$\left(\nabla^2\right)^2 v = 0.$$

5.4   Stokes flow past a sphere

An interesting example is a viscous fluid moving past a sphere. The velocity of the fluid far from the sphere is a constant u. All components of velocity (not just the normal component) vanish at the surface of the sphere.

We are to solve the fourth order partial differential equation $\left(\nabla^2\right)^2 v = 0$ with the boundary conditions that

$$v = 0, \quad \text{for } r = a,$$
$$v_i dx^i \to u dz, \quad \text{for } r \to \infty.$$

Choose a spherical polar coordinate system with origin at the center of the sphere and axis along u. The surface of the sphere is at $r = a$. The metric is

$$ds^2 \equiv g_{ij}dx^i dx^j = dr^2 + r^2 d\theta^2 + r^2 \sin^2\theta \, d\phi^2.$$

We choose $x^1 = r, x^2 = \theta, x^3 = \phi$.

Incompressibility is the condition

$$\text{div } v = 0, \iff \partial_i\left[\sqrt{g}\,g^{ij}v_j\right] = 0.$$

A vector field of the form

$$v = \nabla\alpha \times \nabla\beta,$$

for a pair of scalar fields, solves the incompressibility condition. In tensor notation

$$v_i = \frac{1}{\sqrt{g}}g_{ik}\epsilon^{klm}\partial_l\alpha\,\partial_m\beta.$$

For us

$$\sqrt{g} = r^2\sin\theta,$$

and g_{ij} is diagonal.

We expect the velocity field to have r and θ components but not a ϕ-component: there is a symmetry under rotations around the axis of the system. That is, $v_3 = 0$. This can be accomplished by making the ansatz that $\alpha = \phi$ and that $\frac{\partial\beta}{\partial\phi} = 0$:

$$v_1 = -\frac{1}{r^2\sin\theta}\partial_\theta\beta, \quad v_2 = \frac{1}{\sin\theta}\partial_r\beta.$$

Vorticity, which is the curl of velocity, is given in tensor notation by

$$\omega_i = \frac{1}{\sqrt{g}}g_{ij}\epsilon^{jkl}\partial_k v_l,$$

leading to

$$\omega_1 = \omega_2 = 0,$$

$$\omega_3 = \sin\theta\frac{\partial}{\partial\theta}\left[\frac{1}{r^2\sin\theta}\frac{\partial\beta}{\partial\theta}\right] + \partial_r^2\beta.$$

Then the vorticity is purely angular:

$$\omega = \mathcal{M}\beta\nabla\phi,$$

where

$$\mathcal{M}\beta = \frac{\partial^2 \beta}{\partial r^2} + \frac{\sin\theta}{r^2} \frac{\partial}{\partial\theta} \left(\frac{1}{\sin\theta} \frac{\partial\beta}{\partial\theta} \right).$$

Since

$$\nabla^2 v = -\nabla \times \omega,$$

we get

$$\nabla^2 v = \nabla\alpha \times [\nabla(\mathcal{M}\beta)].$$

Thus the Laplacian on v has the same effect as replacing $\beta \to \mathcal{M}\beta$. So, doing the whole argument twice shows $(\nabla^2)^2 v = 0$ is the same as

$$\mathcal{M}^2\beta = 0.$$

Using $z = r\cos\theta$

$$dz = \cos\theta\, dr - r\sin\theta\, d\theta.$$

The boundary conditions become

$$\partial_\theta\beta = 0 = \partial_r\beta, \quad r = a,$$

$$v_1 = -\frac{1}{r^2 \sin\theta} \partial_\theta\beta \to u\cos\theta, \quad v_2 = \frac{1}{\sin\theta} \partial_r\beta =\to -ur\sin\theta,$$

so that

$$\beta \to -\frac{1}{2} ur^2 \sin^2\theta, \quad \text{for } r \to \infty.$$

We seek a solution of separable form

$$\beta = \sin^2\theta f(r)$$

with

$$f(a) = 0 = f'(a) \quad \text{and} \quad f(r) \to -\frac{1}{2} ur^2, \quad r \to \infty. \tag{5.1}$$

We find

$$\mathcal{M}\beta = \sin^2\theta \mathcal{L}f,$$

where

$$\mathcal{L}f = f'' - \frac{2}{r^2}f.$$

The differential equation is

$$\mathcal{L}^2 f = 0,$$

to be solved with the boundary conditions of eqn (5.1).

Note that

$$\mathcal{L}r^n = [n(n-1) - 2]\, r^{n-2},$$
$$\mathcal{L}^2 r^n = [n(n-1) - 2]\, [(n-2)(n-3) - 2]\, r^{n-4},$$

with roots at $n = -1, 2, 1, 4$. Thus, the solution to the equation is

$$f = \frac{A_{-1}}{r} + A_1 r + A_2 r^2 + A_4 r^4.$$

From the behavior at infinity

$$A_4 = 0, \quad A_2 = -\frac{1}{2}u.$$

Putting in also the b.c. at $r = a$,

$$f(r) = -\frac{1}{4}u\frac{(r-a)^2(2r+a)}{r}.$$

Working our way back

$$v_1 = -\frac{1}{r^2 \sin\theta}\partial_\theta\left[\sin^2\theta f(r)\right] = u\cos\theta\left[1 - \frac{3}{2}\frac{a}{r} + \frac{a^3}{2r^3}\right],$$
$$v_2 = \frac{1}{\sin\theta}\partial_r\beta = \sin\theta f'(r) = -ur\sin\theta\left[1 - \frac{3a}{4r} - \frac{a^3}{4r^3}\right].$$

To compare with the notation of (Landau and Lifshitz, 1969), you must set $v_r = v_1$, $v_\theta = \frac{v_2}{r}$. This is because we define

$$v = v_1\,dr + v_2\,d\theta + v_3\,d\phi,$$

while they have

$$v = v_r e_r + v_\theta e_\theta + v_\phi e_\phi,$$

where the e_r, e_θ, e_ϕ form a triad of covariant vectors of unit length parallel to $dr, d\theta, d\phi$.

From the velocity field we can determine the pressure:

$$\nabla p = \eta \nabla^2 v = -\eta \nabla \alpha \times [\nabla(\mathcal{M}\beta)],$$

$$\partial_i p = -\eta \frac{1}{\sqrt{g}} g_{ij} \epsilon^{jkl} \partial_k \phi \partial_l (\mathcal{M}\beta),$$

$$\partial_r p = \eta \frac{1}{r^2 \sin\theta} \partial_\theta (\mathcal{M}\beta)$$

$$= \eta \frac{1}{r^2 \sin\theta} \partial_\theta \left(\sin^2\theta \mathcal{L}f \right)$$

$$= 2\eta u \cos\theta \frac{1}{r^2} \left[f'' - \frac{2}{r^2}f \right],$$

$$\partial_r p = 3\eta au \cos\theta \frac{1}{r^3},$$

leading to

$$p = p_0 - \frac{3\eta au \cos\theta}{2r^2}.$$

The other quantity of interest is the stress due to viscosity. By integrating this over the surface of the sphere we get the net viscous force on the sphere. The answer turns out to be

$$F = 6\pi \eta au.$$

Recall that this is in the limit of small Reynolds number. It is possible to calculate the corrections as a power series in $\mathcal{R}$, a kind of perturbation theory that takes into account the nonlinear terms:

$$F = 6\pi au\eta \left[1 + \frac{3}{8}\mathcal{R} - \frac{9}{4}\mathcal{R}^2 \log\left(\frac{1}{\mathcal{R}}\right) + \cdots \right].$$

This is reminiscent of "renormalization," an idea first discovered in quantum field theory. It is as though the physical viscosity is replaced by an effective or "renormalized" value that depends on the Reynolds number:

$$\eta_{\text{eff}} = \eta \left[1 + \frac{3}{8}\mathcal{R} - \frac{9}{4}\mathcal{R}^2 \log\left(\frac{1}{\mathcal{R}}\right) + \cdots \right].$$

5.5 Vortex with dissipating core

As long as the vorticity points along the z-axis, and depends only on r,

$$\omega(r, \theta, z, t) \cdot \nabla = h(r)\frac{\partial}{\partial z},$$

we have a static solution of the Euler equations. This is because the velocity is $v \cdot \nabla = f(r)\frac{\partial}{\partial \theta}$ so that $v \cdot \nabla\omega = 0 = \omega \cdot \nabla v$:

$$\frac{\partial \omega}{\partial t} + v \cdot \nabla\omega - \omega \cdot \nabla v = 0.$$

The relation between velocity and vorticity gives

$$\frac{1}{r^2}\partial_r[r^2 f(r)] = h(r).$$

If we include viscosity, the Navier–Stokes equations

$$\frac{\partial \omega}{\partial t} + v \cdot \nabla\omega - \omega \cdot \nabla v = \nu\nabla^2\omega$$

no longer have such a static solution: the vortex core dissipates. The nonlinear terms still vanish because of the cylindrical symmetry. The ansatz

$$\omega(r, \theta, z, t) = h(r, t)\frac{\partial}{\partial z}$$

leads instead to the scalar diffusion equation

$$\frac{\partial h}{\partial t} = \nu\nabla^2 h.$$

A solution is the Gaussian

$$h(r, t) = h_0 \frac{e^{-\frac{r^2}{4\nu t}}}{4\pi\nu t}.$$

This is a vortex whose strength decays in time as $\frac{1}{\nu t}$. Also, the core radius grows as $\sqrt{\nu t}$. This shows that viscosity can also be important away from boundaries. Viscosity drains energy from any region where the velocity gradients are large.

6

Shocks

A good rule in most of physics is to "follow the energy." How different kinds of energy are changed into each other (e.g., kinetic to potential energy or to heat) is key to physical behavior. But sometimes it also pays to "follow the information." How does one part of a system know another part has changed? And when does it know that?

Sound waves carry information about changes of pressure. In an incompressible fluid, this happens instantaneously: the fluid can move around the high pressure regions. A supersonic flow cannot do that and can slam into regions where it is slowed down abruptly. This creates discontinuities (more precisely rapid changes) in velocity and pressure called shocks.

At the opposite extreme from incompressible fluids are those where the speed of sound c_s is zero. Since $c_s^2 = \frac{\partial p}{\partial \rho}$, in this limit the pressure (instead of density) is constant. This could mean that the fluid velocity is much greater than the sound speed (i.e., infinite Mach number). Or, that the collisions between molecules that make up the fluid are so rare (e.g., interstellar medium) that it exists effectively at zero pressure. There are simplifications in this limit as well, which help us to understand shocks.

6.1 The Burgers equations

The simplest case is when space is one-dimensional. In an approximation where we ignore pressure and viscosity, we get

$$\frac{\partial v}{\partial t} + v \frac{\partial v}{\partial x} = 0, \quad \text{Inviscid Burgers equation.} \qquad (6.1)$$

We need to solve this equation given the initial velocity distribution

$$v(x, 0) = u(x).$$

The key to solving this problem is to think in the Lagrangian picture. In the approximation in which we are studying this problem (no dissipation, no pressure), we are ignoring the effects of particle collisions. Each molecule (or fluid element) will move

Fluid Mechanics: A Geometrical Point of View. S. G. Rajeev © S. G. Rajeev 2018.
Published in 2018 by Oxford University Press. DOI: 10.1093/oso/9780198805021.001.0001

along a straight line with the velocity it had at the beginning. If it happened to start at some position X at time $t = 0$, after a time t it will be at $X + u(X)t$. So, the velocity of the fluid at some point (x, t) in space-time is simply the velocity at the point where that fluid element was at the initial time.

We must solve the equation

$$x = X + u(X)t \qquad (6.2)$$

to find $X(x, t)$, that is, the position of the fluid element at the initial time in order that it moves to x in time t. By the inverse function theorem, a unique smooth solution $X(x, t)$ exists as long as the r.h.s. has non-zero derivative. If t is small enough for $1 + u'(X)t$ to be positive, the solution $X(x, t)$ will exist and will be smooth. The first singularity occurs at

$$t_* = \min_X \left[-\frac{1}{u'(X)} \right]_+ ,$$

that is, when $1 + u'(X)t$ first vanishes.

Here $[\cdots]_+$ denotes the positive part:

$$[Y]_+ = \begin{cases} Y & Y > 0 \\ 0 & Y \leq 0 \end{cases} .$$

For $t < t_*$ the solution is

$$v(x, t) = u(X(x, t)).$$

Let us now verify that this satisfies the Burgers equation; that it satisfies the initial condition is obvious. Using implicit differentiation of eqn (6.2) the space and time derivatives are

$$\left[1 + tu'(X(x, t)) \right] \frac{\partial X(x, t)}{\partial x} = 1,$$

$$\left[1 + tu'(X(x, t)) \right] \frac{\partial X(x, t)}{\partial t} + u(X(x, t)) = 0,$$

so that

$$\frac{\partial X(x, t)}{\partial t} + u(X(x, t)) \frac{\partial X(x, t)}{\partial x} = -\frac{u(X(x, t))}{1 + tu'(X(x, t))} + u(X(x, t)) \frac{1}{1 + tu'(X(x, t))} = 0.$$

Moreover,

$$\frac{\partial u(X(x, t))}{\partial t} + u(X(x, t)) \frac{\partial u(X(x, t))}{\partial x} = u'(X(x, t)) \left[\frac{\partial X}{\partial t} + u(X(x, t)) \frac{\partial X}{\partial x} \right] = 0.$$

Once you know v, you can also solve the associated conservation law for mass

$$\frac{\partial \rho}{\partial t} + \frac{\partial}{\partial x} [\rho v] = 0$$

given some initial distribution

$$\rho(x, 0) = \rho_0(x).$$

....

Exercise 6.1 Verify that

$$\rho(x, t) = \frac{1}{1 + u'\left(X(x, t)\right) t} \rho_0\left(X(x, t)\right)$$

satisfies the conservation equation and initial condition.

....

The factor in front is the Jacobian of the transformation $x \mapsto X(x, t)$, that is, the inverse of the derivative of $x(X, t) = X + u(X)t$ w.r.t. X. For $t < t_*$ this is also non-singular. Let us see what happens as we approach the critical time t_* in an example.

Example 6.1

Suppose $u(x) = \frac{1}{2} + \frac{1}{1+x^2}$, $\rho_0(x) = 1$. Then $-\frac{1}{u'(x)} = \frac{(x^2+1)^2}{2x}$. Its minimum is at $x = \frac{1}{\sqrt{3}}$ and so $t_* = \frac{8}{3\sqrt{3}}$. We should find X by solving

$$x = X + \left(\frac{1}{2} + \frac{1}{1 + X^2}\right) t.$$

It can be turned into a cubic for X; when $0 \le t < t_*$ there is a unique real solution and two complex solutions. (It is too messy to display the analytic solution of the cubic; not difficult to find using symbolic calculation though.)

If we plot the solutions for velocity Figures 6.1 and 6.2, we see that as $t \to t_*$ there is shock: a steep drop in velocity. If we start with a uniform initial density and plot the evolved density we will get a sharp peak at the shock. There is a sort of "traffic jam" as particles pile up at the shock. Notice how the density drops below the average value elsewhere so that the total mass is the same.

6.2   The Cole–Hopf transformation

Once the density and velocity gradients become large, it is not correct any more to ignore the terms due to pressure $\frac{\nabla p}{\rho}$ or dissipation $\nu\nabla^2 v$. These effects will be important inside the shock and will also mollify a rapid change of velocity. It would be nice to have

$t = 0, 0.5t_*, 0.9999t_*$

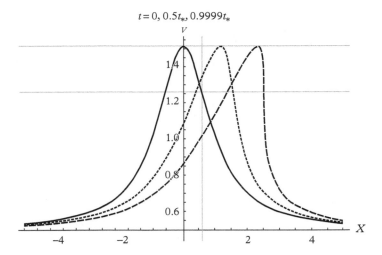

Figure 6.1 *Velocity at a shock*

$t = 0, 0.5t_*, 0.9999t_*$

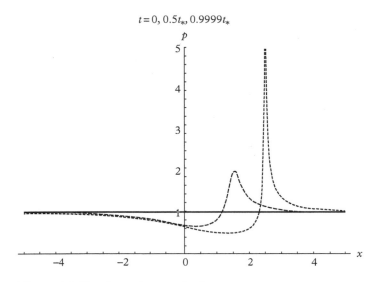

Figure 6.2 *Density at the same shock*

a simple model, that is still exactly solvable, incorporating these effects. This is the point of the Burgers equation:

$$\frac{\partial v}{\partial t} + v \frac{\partial v}{\partial x} = v \frac{\partial^2 v}{\partial x^2}. \quad \text{Burgers equation} \tag{6.3}$$

It still ignores pressure, and treats kinematic viscosity v as a constant. These sacrifices are worth it because by a clever change of variables this can be turned into a linear equation.

To understand this trick, it is useful to rewrite the inviscid Burgers equation in terms of the velocity potential:

$$v = \frac{\partial S}{\partial x},$$

$$\frac{\partial v}{\partial t} + v\frac{\partial v}{\partial x} = \frac{\partial}{\partial x}\left[\frac{\partial S}{\partial t} + \frac{1}{2}\left(\frac{\partial S}{\partial x}\right)^2\right].$$

Thus the Burgers equation becomes

$$\frac{\partial S}{\partial t} + \frac{1}{2}\left(\frac{\partial S}{\partial x}\right)^2 = f(t)$$

for some function $f(t)$ of time alone.

We can make the shift $S \to S + \int^t f(t)dt$ without changing v; then the equation becomes

$$\frac{\partial S}{\partial t} + \frac{1}{2}\left(\frac{\partial S}{\partial x}\right)^2 = 0.$$

You recognize this as the Hamilton–Jacobi equation of classical mechanics, for a hamiltonian $H(x, p) = \frac{p^2}{2}$. That is, a free particle. This should not surprise: the inviscid Burgers equation describes the motion of a collection of free particles, ignoring collision effects such as pressure and viscosity.

How can we modify this to include some dissipation? Recall that viscosity is caused by a diffusion of momentum, due to the random thermal motion of molecules. The Burgers equation can be turned into the diffusion equation by a change of variables.

First, the transformation $v = \frac{\partial S}{\partial x}$ above turns it into

$$\frac{\partial v}{\partial t} + v\frac{\partial v}{\partial x} - \nu\frac{\partial^2 v}{\partial x^2} = \frac{\partial}{\partial x}\left[\frac{\partial S}{\partial t} + \frac{1}{2}\left(\frac{\partial S}{\partial x}\right)^2 - \nu\frac{\partial^2 S}{\partial x^2}\right].$$

That is,

$$\frac{\partial S}{\partial t} + \frac{1}{2}\left(\frac{\partial S}{\partial x}\right)^2 - \nu\frac{\partial^2 S}{\partial x^2} = 0.$$

If you know the connection of the Hamilton–Jacobi equation to the Schrödinger equation you will be motivated to try the ansatz

$$\psi = e^{\alpha S}$$

for a constant a to be determined soon:

$$\frac{\partial \psi}{\partial t} - v\frac{\partial^2 \psi}{\partial x^2} = \psi \left[a\frac{\partial S}{\partial t} - v \left\{ a^2 \left(\frac{\partial S}{\partial x}\right)^2 + a\frac{\partial^2 S}{\partial x^2} \right\} \right].$$

If we choose $-va = \frac{1}{2}$,

$$\frac{\partial \psi}{\partial t} - v\frac{\partial^2 \psi}{\partial x^2} = a\psi \left[\frac{\partial S}{\partial t} + \frac{1}{2}\left(\frac{\partial S}{\partial x}\right)^2 - v\frac{\partial^2 S}{\partial x^2} \right].$$

In summary, the Burgers equation reduces to

$$\frac{\partial \psi}{\partial t} = v\frac{\partial^2 \psi}{\partial x^2}, \quad \text{Diffusion equation} \tag{6.4}$$

under the transformation

$$v = -\frac{1}{2v}\frac{1}{\psi}\frac{\partial \psi}{\partial x}. \quad \text{Cole–Hopf transformation} \tag{6.5}$$

....

Exercise 6.2 Verify that the kernel

$$K(x, y; t) = \frac{e^{-\frac{(x-y)^2}{4vt}}}{\sqrt{4\pi vt}}$$

satisfies the diffusion equation, and that $\int K(x, y; t)dy = 1$. Hence that $\lim_{t \to 0+} K(x, y; t) = \delta(x - y)$.

....

The solution to the diffusion equation (eqn 6.4)given the initial condition

$$\psi(x, 0) = \psi_0(x)$$

is given by integrating over the kernel:

$$\psi(x, t) = \int \frac{e^{-\frac{(x-y)^2}{4vt}}}{\sqrt{4\pi vt}} \psi_0(y)dy.$$

This solves the Burgers equation through the Cole–Hopf transformation.

6.3   The limit of small viscosity

In the limit $\nu \to 0$ we should recover the solution to the inviscid Burgers equation found by the method of characteristics. Setting $\psi = \mathrm{e}^{-\frac{S}{2\nu}}$,

$$\mathrm{e}^{-\frac{S(x,t)}{2\nu}} = \frac{1}{\sqrt{4\pi\nu t}} \int \exp\left[-\frac{1}{2\nu}\left\{\frac{(x-y)^2}{2t} + S_0(y)\right\}\right] dy. \tag{6.6}$$

In the limit of small ν the integral will be dominated by a region near the minimum of the quantity in the curly brackets:

$$\bar{S}(x,t) = \min_y \left\{\frac{(x-y)^2}{2t} + S_0(y)\right\}.$$

This is the "steepest descent" approximation; there is a version of this in every branch of theoretical physics. The minimum must satisfy

$$\frac{y-x}{t} + \frac{\partial S_0(y)}{\partial y} = 0.$$

Or, since $\frac{\partial S_0(y)}{\partial y} = u(y)$,

$$\frac{y-x}{t} + u(y) = 0,$$

$$x = y + u(y)t.$$

This is exactly eqn (6.2), except that what we used to call X is now called y. For $t < t_*$ there is a unique solution. When $t > t_*$ there are multiple solutions. Naively, you might think this leads to a multi-valued solution for velocity and density.

But a careful analysis of what happens beyond $t = t_*$ should use the full integral formula, which holds even when viscosity is non-zero.

6.4   Maxwell–Lax–Oleneik minimum principle

An advantage of reducing the study of shocks in the Burgers equation to the integral formula is that the same formula occurs in many other areas of physics: optics, quantum mechanics, thermodynamics and statistical physics. We can take ideas from these areas and apply them here. Maxwell solved exactly the same problem of multiple minima in the context of thermodynamics. This has now become a general method.

Thinking in terms of S suggests a natural solution to our dilemma when there are several solutions to the equation $x = X + u(X)t$. They are all local extrema; some of them may even be local minima of $\frac{(x-y)^2}{2t} + S_0(y)$. In approximating the integral, the largest contribution will come from the solution that has the smallest value for $\frac{(x-y)^2}{2t} + S_0(y)$. Thus, if there are several solutions, we simply pick the one that is the true global minimum, not just a local minimum. The function $\bar{S}(x, t)$ defined this way will be continuous but may not be differentiable. It is an example of a "weak solution" of a partial differential equation, in this case the Hamilton–Jacobi equation. The value of y where this minimum is achieved is thus the correct choice for the Lagrangian coordinate:

$$\bar{X}(x, t) = \arg\min_y \left\{ \frac{[x - y]^2}{2t} + S_0(y) \right\}.$$

This function will have a discontinuity at the shock. In a typical example, there would be three possible solutions. One is a local maximum and the other two are local minima. Let the arguments of the two minima be labeled $X_-(x, t)$ and $X_+(x, t)$; we choose these labels so that $X_-(x, t) \leq X_+(x, t)$. Each of them is a continuous function of x.

Let $S_-(x, t)$ and $S_+(x, t)$ be the corresponding values of S for those solutions. To the left of the shock, $S_-(x, t) < S_+(x, t)$ so that $\bar{X}(x, t) = X_-(x, t)$. At the shock the values of the potential coincide $S_-(x, t) = S_+(x, t)$. To the right, $S_+(x, t) < S_-(x, t)$ so that the minimizing argument changes: $\bar{X}(x, t) = X_+(x, t)$. Thus, although $X_-(x, t), X_+(x, t)$ are continuous functions, $\bar{X}(x, t)$ has a jump discontinuity at the shock.

This leads to the solution of the inviscid Burgers equation chosen by the limit of small viscosity:

$$\bar{v}(x, t) = u\left(\bar{X}(x, t)\right), \quad \bar{\rho}(x, t) = \frac{1}{1 + tu'\left(\bar{X}(x, t)\right)} \rho_0\left(\bar{X}(x, t)\right).$$

Example 6.2

Let us return to the example $u(x) = \frac{1}{2} + \frac{1}{1+x^2}$ we considered earlier. Clearly $S_0(x) = \frac{x}{2} + \arctan x$. It is not difficult to find $\bar{S}(x, t)$ and $\bar{X}(x, t)$ by numerical minimization (e.g., the function NMinimize in Mathematica comes in handy). As expected, $\bar{S}(x, t)$ is a continuous function with a kink (discontinuity in derivative). And $\bar{X}(x, t)$ has a discontinuity at the same point. The velocity and density can then be calculated as above. We plot the results at $t = 0, \frac{1}{2}t_*, t_*, \frac{3}{2}t_*$ (Figure 6.3).

6.4.1 The Cheshire cat

In *Alice's Adventures in Wonderland* she encounters a cat sitting on a tree. It is grinning, but the cat begins to disappears as she watches. She notices that the instant after the cat has fully disappeared, she could still see its grin. A similar phenomenon is exploited in

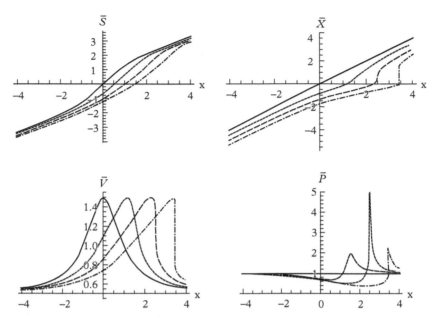

Figure 6.3 *Solution to the Burgers equation in the limit of zero viscosity at* $t = 0, \frac{1}{2}t_*, t_*, \frac{3}{2}t_*$

the yogic exercise of *Thrataka*. Stare at a candle for a while and blow it out. You will still see the flame for a while after it has gone out.

We see something similar in the solution of the inviscid Burgers equation. It should be understood as the limit of the Burgers equation in the limit of small viscosity. As long as the solution to the characteristic equation is unique ($t < t_*$) we do not need to invoke this interpretation. But if there are multiple solutions ($t \geq t_*$) we must appeal to the viscous equation to "break the tie" among them: only the solution that minimizes S can occur as the limit of a solution to the viscous equation. Once we make the proper choice of solution to the characteristic equation, we can continue to ignore dissipation and proceed to solve the inviscid equation. Thus, although the final solution does not depend on viscosity, there is a lingering effect even in the limit as $\nu \to 0$.

6.5 The Riemann problem

We saw that even smooth data can evolve into discontinuities for a compressible ideal gas. So we better be able to solve the initial value problem for discontinuous data as well. The simplest of these is a piecewise constant function. Another reason to look into such initial data is that they are often used to approximate continuous data, for example, when solving equations numerically. Consider a single jump discontinuity

$$u = \begin{cases} u_1 & x < 0 \\ u_2 & 0 < x \end{cases}.$$

We can assign the average value $\frac{u_1+u_2}{2}$ to $u(0)$. Then

$$S_0(x) = \begin{cases} u_1 x & x < 0 \\ u_2 x & 0 < x \end{cases}.$$

The characteristic lines are given by

$$x = \begin{cases} X + u_1 t & X < 0 \\ X + u_2 t & X > 0 \end{cases}.$$

If $u_2 < u_1$ this can be solved exactly, as we explained in the last section.

....

Exercise 6.3 Show that if $u_2 < u_1$ the solution is $\bar{v}(x, t) = \begin{cases} u_1 & x < \frac{u_1+u_2}{2} t \\ u_2 & \frac{u_1+u_2}{2} t < x \end{cases}$. That is, the shock propagates at a speed that is the average of u_1 and u_2.

....

If $u_1 < u_2$ we have a different situation. The solution to the characteristic equation is

$$X(x, t) = \begin{cases} x - u_1 t & x < u_1 t \\ x - u_2 t & u_2 t < x \end{cases}$$

so that

$$v(x, t) = \begin{cases} u_1 & x < u_1 t \\ u_2 & u_2 t < x \end{cases}.$$

But when $u_1 t < x < u_2 t$ there is no solution: there is simply no Lagrange co-ordinate X corresponding to (x, t).

Do we again need to look to viscosity to rescue us? No. The inviscid Burgers equation has a scale invariant solution

$$v(x, t) = \frac{x}{t}.$$

You can check that this provides a continuous interpolation between the solutions in the two known regions:

$$v(x, t) = \begin{cases} u_1 & x < u_1 t \\ \frac{x}{t} & u_1 t < x < u_2 t \\ u_2 & u_2 t < x \end{cases}.$$

So we have found a solution to the inviscid Burgers equation which satisfies the given initial condition.

But how to justify this trick? The discontinuous velocity profile can be thought as the limiting case of a smoothly increasing function. For example,

$$u_\epsilon(x) = \frac{u_1 + u_2}{2} + \frac{u_2 - u_1}{2} \tanh\left[\frac{x}{\epsilon}\right]$$

for some small ϵ. Its integral is

$$S_{0,\epsilon}(x) = \frac{u_1 + u_2}{2} x + \epsilon \frac{u_2 - u_1}{2} \log \cosh\left[\frac{x}{\epsilon}\right]$$

The characteristic equation has a solution now (but it can be only found numerically). Since the initial profile is increasing, there are no shocks; the problem is well-posed. Having found the solution, we can take the limit as $\epsilon \to 0$.

....

Exercise 6.4 Choose $u_1 = 0, u_2 = 1$. Solve the initial value problem numerically for several small values of $\epsilon = 10^{-2}, 10^{-3}, 10^{-6}$. Verify that as $\epsilon \to 0$ the solution tends to the linear interpolation above.

....

7

Boundary Layers

If you set the velocity at the boundary (assumed to be fixed) to zero, the solution to Euler's equations vanishes everywhere. To get non-trivial solutions, only the normal component of velocity should vanish at a boundary. If the tangential velocity not zero, there will be great deal of friction as the fluid brushes past the boundary. Certainly it is wrong to ignore viscosity when there is a large gradient in velocity. So near the boundary, Euler's equations can't be right physically.

Prandtl proposed that the flow be thought of as two separate parts: a thin "boundary layer" (Schlitning, 1979; Rosenhead, 1988) where viscosity matters and a bulk where the fluid flow is ideal (satisfies Euler equations). This is useful in laminar (i.e., not turbulent) flow. When the flow is turbulent, the velocity has large gradients everywhere. Viscosity will dissipate energy and momentum to small scales everywhere.

7.1 Prandtl's theory

Consider a viscous fluid flowing past a plate located along the half-plane $x > 0, y = 0$. The boundary condition is that as $|y| \to \infty$ the fluid flows in the x-direction with a velocity $U(x)$. The latter should be the solution of the ideal fluid equation (the "outer flow," where viscosity is negligible) with appropriate boundary conditions. At the plate, velocity must be zero. The whole problem is homogenous in the z-coordinate. So we make the ansatz that the velocity in the z-direction is zero and that the remaining components of velocity are independent of z. Also, it is static. (There are flows where these conditions are satisfied experimentally.)

Then, the Navier–Stokes equations reduce to two dimensions:

$$\partial_x v_1 + \partial_y v_2 = 0,$$

$$\left[v_1 \partial_x + v_2 \partial_y \right] v_1 = \nu \left[\partial_x^2 + \partial_y^2 \right] v_1 - \partial_x \psi,$$

$$\left[v_1 \partial_x + v_2 \partial_y \right] v_2 = \nu \left[\partial_x^2 + \partial_y^2 \right] v_2 - \partial_y \psi.$$

v_1, v_2 are the x, y components of velocity. ψ is pressure divided by density.

Fluid Mechanics: A Geometrical Point of View. S. G. Rajeev © S. G. Rajeev 2018.
Published in 2018 by Oxford University Press. DOI: 10.1093/oso/9780198805021.001.0001

We can solve the condition for incompressibility by introducing a velocity potential (also called stream function)

$$v_1 = \phi_y, \quad v_2 = -\phi_x.$$

The subscript here means differentiation: $\phi_x \equiv \partial_x \phi$ etc, a convention we adopt from analysts who study partial differential equations.

That still leaves us with two nonlinear equations for two scalar unknowns:

$$\phi_y \phi_{xy} - \phi_x \phi_{yy} = \nu \left[\phi_{xxy} + \phi_{yyy} \right] - \partial_x \psi, \tag{7.1}$$

$$\phi_y \phi_{xx} - \phi_x \phi_{xy} = \nu \left[\phi_{xxx} + \phi_{xyy} \right] + \partial_y \psi. \tag{7.2}$$

It is reasonable to suppose that the derivative of ϕ along the y direction is large compared to that along the x-direction. In other words, that $|v_2| \ll |v_1|$ and the velocity gradients normal to the boundary are large compared to those tangential.

Then eqn (7.1) becomes

$$\phi_y \phi_{xy} - \phi_x \phi_{yy} = \nu \phi_{yyy} - \partial_x \psi.$$

Although ϕ_x is small, it is multiplied by the large quantity ϕ_{yy} so we keep it.

In eqn (7.2), all the terms involving ϕ are small. So we conclude

$$\partial_y \psi = 0.$$

Since ψ depends on x alone, it is determined by its value as $|y| \to \infty$. In this limit $v_2 \to 0$ and $v_1 \to U$. Thus we find the pressure in terms of the outer flow (invoking Bernoulli):

$$\partial_x \psi = -\partial_x \left(\frac{1}{2} U^2 \right).$$

Thus,

$$\phi_y \phi_{xy} - \phi_x \phi_{yy} = \nu \phi_{yyy} + \partial_x \left(\frac{1}{2} U^2 \right).$$

7.2   The Blasius reduction

The simplest case is when the outer flow is a constant U:

$$\phi_y \phi_{xy} - \phi_x \phi_{yy} = \nu \phi_{yyy}. \tag{7.3}$$

The constant U still appears in the bulk condition:

$$\phi(x,y) \to Uy, \quad y \to \infty.$$

In addition, $v_1 = v_2 = 0$ at $y = 0$:

$$\phi_x = 0 = \phi_y, \quad y = 0.$$

The reduced equation is invariant under a scaling where x and y are assigned different weights. This is reasonable, as the rates change along them are quite different by our assumptions. So we must do dimensional analysis treating x and y as independent dimensions. From the boundary condition,

$$[\phi] = [y] + [U].$$

From the Prandtl equation,

$$2[\phi] - 2[y] - [x] = [\phi] - 3[y] + [v].$$

Eliminating $[\phi]$

$$\implies [y] - \frac{1}{2}[x] - \frac{1}{2}[v] + \frac{1}{2}[U] = 0$$

and eliminating y

$$[\phi] - \frac{1}{2}[v] - \frac{1}{2}[U] - \frac{1}{2}[x] = 0.$$

So we define the dimensionless variables

$$\eta = y\sqrt{\frac{U}{vx}}, \quad f = \frac{\phi}{\sqrt{vUx}}.$$

If the solution is scale invariant, f will only depend on the dimensionless combination η. This leads to the Blasius ansatz

$$\phi(x,y) = \sqrt{vUx}f(\eta).$$

The Prandtl equation will reduce to an ordinary differential equation (ODE) for f.

(You could set $v = U = 1$ temporarily to simplify calculations and restore them at the end using dimensional analysis. Or, keep track of all the factors of U, v as you go.)

$$\eta_y = \sqrt{\frac{U}{v}}\frac{1}{\sqrt{x}}, \quad \eta_x = -\sqrt{\frac{U}{v}}\frac{\eta}{2x},$$

$$v_1 = \phi_y = Uf', \quad -v_2 = \phi_x = \sqrt{vU}\left[\frac{f}{2\sqrt{x}} - \frac{\eta}{2\sqrt{x}}f'\right],$$

$$\phi_{xy} = -U\frac{\eta}{2x}f'', \quad \phi_{yy} = \frac{U^{\frac{3}{2}}f''}{\sqrt{v}\sqrt{x}},$$

$$\phi_{yyy} = \frac{U^2}{v}\frac{f'''}{x},$$

so that

$$-\frac{1}{2x}U^2\left[f'\eta f''\right] - \frac{1}{2\sqrt{x}}U^2\left[f - \eta f'\right]\frac{f''}{\sqrt{x}} = U^2\frac{f'''}{x},$$

which reduces to the Blasius equation

$$f''' + \frac{1}{2}ff'' = 0. \tag{7.4}$$

The boundary conditions are: at $y = 0$, $v_1 = v_2 = 0$,

$$f(0) = f'(0) = 0. \tag{7.5}$$

The bulk condition $y \to \infty$, $v_1 \to U$ gives

$$f'(\infty) = 1. \tag{7.6}$$

Although we have the right number of boundary conditions, it is awkward to have them both at infinity and the origin. "Shooting methods" are often unstable and are best avoided, even if you could get away with it here.

Define instead another function satisfying the same ODE (eqn 7.4) but with just initial conditions:

$$g''' + \frac{1}{2}gg'' = 0, \tag{7.7}$$

$$g(0) = 0 = g'(0), \quad g''(0) = 1. \tag{7.8}$$

Then

$$f(\eta) = ag(a\eta)$$

satisfies the ODE (eqn 7.4) and boundary conditions (eqns 7.5 and 7.6) provided that

$$a = \frac{1}{\sqrt{g'(\infty)}}.$$

It is easy enough to solve the initial value problem for g numerically (e.g., in Mathematica). We get a from the asymptotic value of the derivative

$$a = 0.692475$$

and the solution for f can be found easily. The tangential velocity, which will start at zero near the boundary and rise to the bulk value, can now be plotted as in Figure 7.1. It is in remarkable agreement with experiment if the Reynolds number is not too large: see (Tropea et al., 2007, 63).

Experiments show that for large Reynolds numbers $R = \frac{Ux}{\nu} \approx 3 \times 10^5$, the theory fails: the layer becomes turbulent. But the transition to turbulence can be delayed by making the flow free from disturbances, to $R = 3 \times 10^6$. This intermediate region could be like a glass, with many meta-stable solutions.

....

Exercise 7.1 Find the ODE and b.c. for a scale invariant solution to the Prandtl equations (7.1, 7.2) where asymptotically $U \to U_0 x^n$ for some constant exponent n. These solutions describe a boundary layer near a corner whose angle is $\frac{\pi}{n}$. Verify that the special case $n = 0$ reduces to the Blasius solution. Find the numerical solution and plot it for the special case $n = 1$.

....

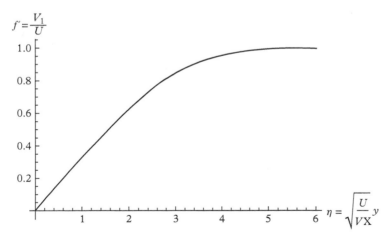

Figure 7.1 *The Blasius solution*

7.3 Weyl's method

(Weyl, 1942) proposed an analytical method to find successively better approximations for g above. Equation (7.7) can be written as the pair

$$h = g'', \quad h' + \frac{1}{2}gh = 0.$$

Using the b.c. $h(0) = g''(0) = 1$ we can solve for h in terms of g:

$$h(y) = e^{-\frac{1}{2}\int_0^y g(y_1)dy_1}.$$

Integrating by parts twice, and using the b.c. of eqn (7.8),

$$\int_0^y g(y_1) \times 1 \times dy_1 = \int_0^y g'(y_1)(y - y_1)dy_1 = \frac{1}{2}\int_0^y h(y_1)(y - y_1)^2 dy_1.$$

Thus the Blasius equation is equivalent to the integral equation

$$h(y) = e^{-\frac{1}{4}\int_0^y (y-y_1)^2 h(y_1)dy_1}.$$

This is the fixed point of the map of functions on the half-line $\psi \mapsto \Phi[\psi]$:

$$\Phi[\psi](y) = e^{-\frac{1}{4}\int_0^y (y-y_1)^2 \psi(y_1)dy_1}.$$

Weyl proved that the sequence

$$h_n = \Phi(h_{n-1}), \quad h_0(y) = 1$$

converges to the solution we seek.
Then

$$h_1(y) = e^{-\frac{y^3}{12}}.$$

We can get the constant we need in this approximation:

$$g_1(\infty) = \int_0^\infty h_1(y)dy = \left(\frac{2}{3}\right)^{2/3} \Gamma\left(\frac{1}{3}\right),$$

$$a_1 = \frac{1}{\sqrt{g_1'(\infty)}} = \frac{\sqrt[3]{\frac{3}{2}}}{\sqrt{\Gamma\left(\frac{1}{3}\right)}} \approx 0.699384,$$

which is close to the numerical value we obtained earlier.

Then

$$f_1(x) = a_1 g_1(a_1 y),$$

where

$$g_1(y) = \int_0^y [y - y_1] h_1(y_1) dy_1$$

$$= \frac{y^2 E_{\frac{1}{3}}\left(\frac{y^3}{12}\right)}{3} - \frac{y^2 E_{\frac{2}{3}}\left(\frac{y^3}{12}\right)}{3} + \left(\frac{2}{3}\right)^{2/3} y\Gamma\left(\frac{1}{3}\right) - 2\sqrt[3]{\frac{2}{3}}\Gamma\left(\frac{2}{3}\right).$$

Here, the exponential integral is defined by

$$E_n(x) = \int_0^\infty \frac{e^{-tx}}{t^n} dt, \quad n > 1.$$

For other values of n it is defined by an analytic continuation.

We can thus calculate the first approximation to the Blasius function this way analytically in terms of exponential integrals:

$$f_1(y) = \frac{y^2 E_{\frac{1}{3}}\left(\frac{y^3}{8\Gamma\left(\frac{1}{3}\right)^{3/2}}\right) - y^2 E_{\frac{2}{3}}\left(\frac{y^3}{8\Gamma\left(\frac{1}{3}\right)^{3/2}}\right) + 2y\Gamma\left(\frac{1}{3}\right)^{3/2} - 4\Gamma\left(\frac{1}{3}\right)\Gamma\left(\frac{2}{3}\right)}{2\Gamma\left(\frac{1}{3}\right)^{3/2}}.$$

The agreement with the numerical solution and this analytical approximation is impressive: the error is never more than 3%. The flat plate is the "hydrogen atom" of boundary layer theories: the simplest case that can be understood analytically.

....

Exercise 7.2 Compare the Weyl's first order approximation to the solution of Blasius equation to a numerical solution.

....

7.4   Drag on a flat plate

The drag on a flat plate facing the flow at zero incidence angle can now be calculated. The viscous stress at the surface is proportional to the normal velocity gradient:

$$\tau_{12}(x, y) = -\nu\rho\left[\partial_1 v_2 + \partial_2 v_1\right] = \nu\rho\left[\phi_{xx} - \phi_{yy}\right].$$

Putting in

$$\phi(x, y) = \sqrt{\nu U x} f\left(y\sqrt{\frac{U}{\nu x}}\right)$$

and setting $y = 0$ we get (recall that $f(0) = 0$)

$$\tau_{12}(x, 0) = -\rho\sqrt{\frac{U^3\nu}{x}} f''(0) = -a^3\rho\sqrt{\frac{U^3\nu}{x}},$$

where $a \approx 0.7$ is the constant we found earlier. Thus the force of a plate of length l and breadth b is (since it has two sides we must multiply by two)

$$F = 2b\int_0^l \tau_{12}dx = -4a^3 b\rho\sqrt{U^3\nu l}.$$

This can be conveniently expressed in terms of a dimensionless quantity, the drag coefficient c_D defined by

$$F = -bl\left[\frac{1}{2}\rho U^2\right]c_D.$$

It makes sense that c_D should depend only on the Reynolds number:

$$c_D = \frac{8a^3}{\sqrt{\mathcal{R}}} \approx \frac{2.6}{\sqrt{\mathcal{R}}}, \quad \mathcal{R} = \frac{Ul}{\nu}.$$

We just resolved d'Alembert's paradox. An infinitesimally thin plate placed at zero incidence angle has zero drag force in the ideal theory (the limit as $\mathcal{R} \to \infty$). Not so for finite $\mathcal{R}$, taking viscosity into account.

7.5   Limitations of boundary layer theory

As with all approximation methods, boundary layer theory has its limits. If the boundary is curved and too long, there will be a point which the layer separates from it. Beyond that point, the flow can even run contrary to the main direction. Another complication is that the boundary gets too long, the Reynolds number grows and the flow within the boundary layer can become turbulent. These are complex phenomena beyond the scope of this book and subject to intense research activity.

It has been established rigorously (Gerard-Varet and Dormy, 2010) that Prandtl's equations are ill-posed for long boundaries. These singularities are presumably the mathematical manifestation of boundary layer separation. There is even a clear understanding of the mechanism of the instability.

8

Instabilities

The unpredictability of fluid flow is to a large extent due to instabilities. In many situations, there is a simple solution (steady flow, or with a periodic time dependence) that is unstable to small perturbations. A first stab at understanding these instabilities is to study the linearization of the Navier–Stokes equations around the simple solution, leading to an eigenvalue problem. A deeper study would require including the non-linearities. To make the problem tractable, some other approximations (that cut down on the number of degrees of freedom) will have to be made. Perhaps this is one route to a theory of turbulence. We study just a couple of illustrative examples, instead of giving a comprehensive list.

8.1   The Rayleigh–Taylor instability

We already studied in Section 4.8 the waves at the interface of water and air, caused by gravity. For small amplitudes and large depths, we found a dispersion relation $\omega = \sqrt{gk}$. Here the heavier fluid (water) is below the lighter one (air is so light its inertia was ignored). The opposite situation, when a heavier fluid is on top of a lighter fluid, would be an unstable arrangement: it would be as though we reversed the sign of g in the above dispersion relation, causing the frequency to be become imaginary. We will look into this in more detail.

Consider the interface between two infinitely thick fluid layers, the upper one of density ρ_1 and the lower one of density ρ_2. We begin by ignoring surface tension and viscosity, to keep matters simple. Surface tension will modify the behavior at small wavenumbers and viscosity matters more at high wavenumbers. These modifications can be can be put in later.

We assume that the flow of both fluids is incompressible: the fluid velocities are small compared to that of sound. (There are situations in astrophysics, for example, where supersonic Rayleigh–Taylor instability instabilities occur. Once you understand the basic situation you can figure them out too.) The effect of vorticity can be ignored as well for now. Thus we have a velocity potential ϕ with

$$v = \nabla\phi,$$
$$\nabla \cdot v = \nabla^2\phi = 0.$$

Fluid Mechanics: A Geometrical Point of View. S. G. Rajeev © S. G. Rajeev 2018.
Published in 2018 by Oxford University Press. DOI: 10.1093/oso/9780198805021.001.0001

At equilibrium, the interface between the two liquids is a plane, which we chose to be $z = 0$.

The departure from equilibrium is a deformation of this interface to some surface $\zeta(x, y, t)$. The boundary condition is that the normal velocity of the fluids be equal to the velocity of the interface:

$$\frac{\partial \phi}{\partial z} = \frac{D\zeta}{dt} \equiv \frac{\partial \zeta}{\partial t} + \frac{\partial \phi}{\partial x}\frac{\partial \zeta}{\partial x} + \frac{\partial \phi}{\partial y}\frac{\partial \zeta}{\partial y}.$$

In the absence of surface tension and viscosity, the pressure must be continuous across the interface. Applying Bernoulli's theorem, this is the continuity of

$$\rho \left[\frac{1}{2}(\nabla \phi)^2 + \frac{\partial \phi}{\partial t} + gz \right].$$

If the departure from equilibrium is small, we can linearize these equations. We seek a solution with exponential dependence on x, y, t. That is,

$$\phi(x, y, z, t) = \Phi(z)e^{ik_1 x + ik_2 y + \gamma t},$$
$$\zeta(x, y, t) = Ze^{ik_1 x + ik_2 y + \gamma t}.$$

This is a wave on the surface with wavenumbers k_1, k_2 in the x, y directions. The dependence of ϕ on the z-coordinate is determined by the Laplace equation and the b.c. that the fluid be at rest far from the surface:

$$\phi(x, y, z, t) = \begin{cases} Ae^{-kz}e^{ik_1 x + ik_2 y + \gamma t} & z > 0 \\ Be^{kz}e^{ik_1 x + ik_2 y + \gamma t} & z < 0 \end{cases}$$

with

$$k = \sqrt{k_1^2 + k_2^2}.$$

The constants A, B, Z are determined by the conditions across the interface. Requiring $\frac{\partial \phi}{\partial z} = \frac{\partial \zeta}{\partial t}$ (ignoring the nonlinear terms) as $z \to 0$ from above and below gives the two conditions

$$-kA = \gamma Z = kB.$$

The continuity of $\rho \left[\frac{\partial \phi}{\partial t} + gz \right]$ (again ignoring the nonlinear term) gives

$$\rho_1 [\gamma A + gZ] = \rho_2 [\gamma B + gZ].$$

A little algebra eliminates A, B, Z in the above three equations to give a dispersion relation:

$$\gamma = \pm \sqrt{g \frac{\rho_1 - \rho_2}{\rho_1 + \rho_2} k}.$$

If this quantity is real, the time dependence is exponential (unstable equilibrium). If γ is imaginary, the solution is oscillatory (stable). Clearly, the situation with the denser fluid on top, $\rho_1 > \rho_2$, is unstable. Note that the higher the wavenumber, the greater the instability. If you include the effect of viscosity, γ as a function of k reaches a maximum value and then decreases. If you include surface tension, there is a critical value of k below which the interface is stable. You can spend your life pursuing the various subplots of this story.

- If the whole fluid system is placed on an oscillating support, the Rayleigh–Taylor (RT) instability instability is inhibited (Wolf, 1969). This is similar to the Kapitza pendulum, which is able to "stand on its head" (see Section 8.8).

- If the density varies continuously (instead of having just two values) the stable situation is for it to decrease with height. It is possible to derive an eigenvalue problem for infinitesimal perturbations of a density profile. The limiting case of a sudden change reduces to the above analysis (Chandrasekhar, 1961).

- In a supernova, a dense spherical layer of nuclear matter is accelerated into the thinner outer atmosphere of the star. This acceleration is equivalent to a radial gravitational field (equivalence principle) and the situation is similar to the above, with a heavy fluid being pushed into a lighter fluid. The RT instability causes the heavy fluid to be broken up into tiny "fingers" which are visible in photographs of the remnants.

- Inertial confinement of fusion is an inversion of the above picture of a supernova. A spherical pellet with a dense outer layer and a thin inner layer is bombarded with laser light from the outside. Some of the outer layer is ablated away. The recoil causes the denser fluid to be pushed inward, compressing the inner layer (which contains deuterium). Here the RT instability is a hindrance: instead of compressing the inner layer uniformly, the density variation breaks up into irregular "fingers." Is there a way to inhibit the RT instability instability of inertial confinement fusion by shining a rapidly varying laser field on the target, instead of shaking the fluid as in (Wolf, 1969)?

Exercise 8.1 Find the dispersion relation of the instability when the lower fluid is of finite depth d and the upper fluid is infinitely thick.

Exercise 8.2 Reseach project Incorporate surface tension and viscosity into the above analysis, staying within the linear approximation.

8.2   Linearization of Navier–Stokes equations

The traditional approach to studying instabilities starts with infinitesimal perturbations: the linearization of the Navier-Stokes (NS) equations around some given solution V. In the last example, this unperturbed solution was simply $V = 0$. Now we consider a non-zero V, but keep it quite simple. It depends only on one variable and has only one non-zero component.

For small enough Reynolds number, there are small oscillations; above some critical Reynolds number $\mathcal{R}_c$, some of these frequencies become imaginary: the signature of an instability.

There are situations where a linear analysis indicates stability, yet the system is actually unstable even against small disturbances. Even so, the linearization needs to be understood first.

8.2.1   Poiseuille and plane Couette flows

Perhaps the simplest solution to the NS equations is a velocity field that has just one non-zero Cartesian component (say the x-direction); and it depends only one other Cartesian coordinate (say y). An example of this is plane Poiseuille flow: a fluid moves between two horizontal plates located at $y = -a$ and $y = a$. There is a constant pressure gradient $\partial_x p$ along the x-axis that pushes the fluid along at a constant flow rate. There is a solution to the NS equation with velocity pointed along the x-axis and depending only on the y-axis.

The NS equation in this case reduces to

$$\nu \frac{d^2}{dy^2} V(y) = \partial_x p, \quad V(-a) = 0 = V(a).$$

The solution is a parabolic velocity profile:

$$V(y) = V_0 \left[1 - \left(\frac{y}{a} \right)^2 \right].$$

V_0 is the velocity at the center $y = 0$ of the channel. Pressure is a linear function of x.

Another example is plane Couette flow. Here again, the fluid moves between two parallel plates, but there is no pressure gradient. Instead, one of the bounding plates is moving at a constant velocity V_0. The NS equation becomes

$$\frac{d^2}{dy^2} V(y) = 0, \quad V(-a) = 0, \quad V(a) = V_0$$

with a linear solution

$$V(y) = \frac{1}{2}(a + y)V_0.$$

8.3 Orr–Sommerfeld equation

Let us consider a small perturbation to NS equations around a solution of the type described above:

$$v(x, y, z, t) = (V(y) + u_1(x, y, z, t), u_2(x, y, z, t), u_3(x, y, z, t)).$$

To first order,

$$\frac{\partial u_1}{\partial t} + V(y)\frac{\partial u_1}{\partial x} + u_2\frac{\partial V}{\partial y} = \nu\nabla^2 u_1 - \frac{\partial p}{\partial x},$$

$$\frac{\partial u_2}{\partial t} + V(y)\frac{\partial u_2}{\partial x} = \nu\nabla^2 u_2 - \frac{\partial p}{\partial y},$$

$$\frac{\partial u_3}{\partial t} + V(y)\frac{\partial u_3}{\partial x} = \nu\nabla^2 u_3 - \frac{\partial p}{\partial z},$$

along with incompressibility

$$\nabla \cdot u = 0.$$

To eliminate u_1, u_3, p, differentiate the first of the equations w.r.t. x_1 above, the third equation w.r.t. x_3, add them and use the fourth equation:

$$-\frac{\partial}{\partial t}\partial_y u_2 - V(y)\frac{\partial}{\partial x}\partial_y u_2 + \partial_x u_2\frac{\partial V}{\partial y} = -\nu\nabla^2\partial_y u_2 - \left[\frac{\partial^2 p}{\partial x^2} + \frac{\partial^2 p}{\partial z^2}\right].$$

Thus eliminating u_1, u_3, differentiate yet again w.r.t. y to get

$$-\frac{\partial}{\partial t}\partial_y^2 u_2 - V(y)\frac{\partial}{\partial x}\partial_y^2 u_2 + \partial_x u_2\frac{\partial^2 V}{\partial y^2} = -\nu\nabla^2\partial_y^2 u_2 - \left[\frac{\partial^2}{\partial x^2} + \frac{\partial^2}{\partial z^2}\right]\partial_y p.$$

Apply the operator $\frac{\partial^2}{\partial x^2} + \frac{\partial^2}{\partial z^2}$ to the second component of the linearized NS to get

$$\frac{\partial}{\partial t}\left(\partial_x^2 u_2 + \partial_z^2 u_2\right) + V(y)\frac{\partial}{\partial x}\left(\partial_x^2 u_2 + \partial_z^2 u_2\right) = \nu\nabla^2\left(\partial_x^2 u_2 + \partial_z^2 u_2\right) - \left[\frac{\partial^2}{\partial x^2} + \frac{\partial^2}{\partial z^2}\right]\frac{\partial p}{\partial y}.$$

Subtracting the last two eliminates p:

$$\frac{\partial}{\partial t}\nabla^2 u_2 + V(y)\frac{\partial}{\partial x}\nabla^2 u_2 - V''(y)\partial_x u_2 = \nu\nabla^4 u_2.$$

Note that these equations are translation invariant in x, z, t. This suggests the ansatz

$$u_2(x, y, z, t) = \text{Re } U(y)e^{i[k_1 x + k_3 z - \omega t]}$$

for a complex-valued function $U(y)$.

$$\nabla^2 u_2 = \text{Re}\left[\left\{\frac{d^2}{dy^2} - k^2\right\} U\right] e^{i[k_1 x + k_3 z - \omega t]}, \quad k^2 \equiv k_1^2 + k_2^2,$$

$$\nabla^4 u_2 = \text{Re}\left[\left\{\frac{d^2}{dy^2} - k^2\right\}^2 U\right] e^{i[k_1 x + k_3 z - \omega t]},$$

$$ik_1\left[\left\{V(y) - \frac{\omega}{k_1}\right\}\left\{\frac{d^2}{dy^2} - k^2\right\} U - V''(y)U\right] = \nu\left\{\frac{d^2}{dy^2} - k^2\right\}^2 U.$$

This ODE is the Orr–Sommerfeld equation. The parameters from the boundary of the problem V_0, a define the Reynolds number:

$$\mathcal{R} = \frac{V_0 a}{\nu}.$$

A cheap trick to pass to dimensionless variables is to set $V_0 = 1, a = 1, \nu = \frac{1}{\mathcal{R}}$, giving

$$\left\{V(y) - \frac{\omega}{k_1}\right\}\left\{\frac{d^2}{dy^2} - k^2\right\} U - V''(y)U = \frac{1}{ik_1 \mathcal{R}}\left\{\frac{d^2}{dy^2} - k^2\right\}^2 U.$$

For simplicity we can consider the special case $k_3 = 0$ so that $k_1 = k$. Set $c = \frac{\omega}{k}$. For real ω, k the perturbation will propagate as a wave with phase velocity c. We get an eigenvalue problem for c given $\mathcal{R}$ and k:

$$ik\mathcal{R}\left[\{V(y) - c\}\left\{\frac{d^2}{dy^2} - k^2\right\} U - V''(y)U\right] = \left\{\frac{d^2}{dy^2} - k^2\right\}^2 U.$$

In dimensionless variables, Poiseuille flow has

$$V(y) = \left[1 - y^2\right], \quad U(\pm 1) = 0 = U''(\pm 1).$$

For planar Couette flow,

$$V(y) = \frac{1 + y}{2}$$

with the same boundary conditions.

When the eigenvalue c has a positive imaginary part for real k we have an instability (because ω would also have a positive imaginary part and the disturbance would grow with time.)

Among the people who have studied this are Heisenberg,[1] (Chandrasekhar, 1961) and (Orszag, 1971). The smallest value of the Reynolds number for which ω has an imaginary part is the linear theory's prediction for the critical Reynolds number $\mathcal{R}_c$ at which turbulence sets in. For the planar Poiseuille flow (Orszag, 1971), $R_c \approx 5700$. There is the possibility that small (but not infinitesimal) disturbances can grow even for $\mathcal{R} < \mathcal{R}_c$. This complicates the story.

Later, in Section 13.9, we will solve numerically a version of the Orr–Sommerfeld equation, to find its prediction of the critical Reynolds number.

8.3.1 Linear stability of pipe Poiseuile flow

Experimentally, it is simpler to study the flow of a liquid through a circular pipe. There is a long history of studying the instabilities of pipe Poiseuille flow, both mathematically and experimentally. The conclusion is that (contrary to experimental results) the spectrum of linear perturbations is stable. This puzzle is only now starting to be resolved. It has to do with transient solutions to the linear problem, which grow for a while in time before dying out again. In a strictly linear theory, therefore, they are not instabilities. But if the linear theory is only an approximation to a nonlinear problem (as is usually the case in physics) when they grow to a large (but finite value) some nonlinear effects take over, making the system unstable.

Let us begin with the laminar solution whose stability we will investigate. A fluid moves down an infinitely long circular pipe of radius a. Choose the x-axis to be along the pipe. The NS equations are satisfied by (We looked at this in Section 5.1. The notation here is a bit different: for example, the flow is along the x-axis instead of the z-axis.)

$$V = \left(V_0 \left[1 - \left(\frac{y^2 + z^2}{a^2} \right) \right], 0, 0 \right).$$

On the cylinder $y^2 + z^2 = a^2$, the velocity vanishes, which is the "no-slip" boundary condition.

Since the only non-zero component is along the x-direction and the x-derivatives are zero, we have

$$V \cdot \nabla V = 0$$

and

$$\nabla \cdot V = 0.$$

---

[1] There is an amusing story about Heisenberg's PhD defense (Cassidy, 1993).

And of course $\frac{\partial V}{\partial t} = 0$. So the NS equations reduce to

$$0 = \nu \left[\frac{\partial^2}{\partial y^2} + \frac{\partial^2}{\partial z^2} \right] V - \nabla p.$$

Since the y, z components of V are zero, this says that the pressure per density Ψ is independent of y, z. So the NS equations are satisfied with a constant pressure gradient along the x-axis:

$$\frac{\partial p}{\partial x} = -\frac{4 V_0 \nu}{a^2}.$$

The Reynolds number of the flow is

$$\mathcal{R} = \frac{V_0 a}{\nu}.$$

We can now choose units (i.e., rescale variables) so that

$$a = 1, \quad V_0 = 1.$$

The equation will depend on V_0, a, ν only through the combination $\mathcal{R}$. Then

$$V = \left(1 - y^2 - z^2, 0, 0 \right).$$

Then the NS equations become

$$\frac{\partial v}{\partial t} + v \cdot \nabla v = \frac{1}{\mathcal{R}} \nabla^2 v - \nabla p, \quad \nabla \cdot v = 0.$$

Consider solutions infinitesimally close to V:

$$v = V + u,$$
$$p = P + p_1.$$

Then the linearized NS equations are

$$\frac{\partial u}{\partial t} + V \cdot \nabla u + u \cdot \nabla V = \frac{1}{R} \nabla^2 u - \nabla p_1, \quad \nabla \cdot v = 0,$$

with the b.c.

$$u = 0, \quad \text{for } y^2 + z^2 = 1.$$

More explicitly,

$$\frac{\partial u}{\partial t} + \left[1 - y^2 - z^2\right]\frac{\partial u}{\partial x} - \begin{pmatrix} 2yu_2 + 2zu_3 \\ 0 \\ 0 \end{pmatrix} = \frac{1}{R}\nabla^2 u - \nabla p_1.$$

If we do a Fourier transform in x, t and a Fourier series in the angular coordinate, this can be turned into a system of ODEs. The corresponding eigenvalue equation can be solved numerically. The conclusion is that there is no instability even for large R. This conflicts with observation, where turbulence kicks in around $\mathcal{R} \sim 3000$.

8.4   Transient solutions of linear equations

The validity of the linear approximation needs to be re-examined (Trefethen et al., 1993). Suppose we are solving a system of linear ODEs,

$$\frac{du}{dt} = Au,$$

for some constant matrix A. The solution is

$$u(t) = e^{tA}u(0),$$

where the matrix exponential is defined by the usual infinite series

$$e^A = \sum_{k=0}^{\infty} \frac{A^k}{k!}.$$

How big does the solution get? The norm of a vector is defined by

$$|u| = \sqrt{u^\dagger u},$$
$$|u(t)|^2 = u^\dagger(0)e^{tA^\dagger}e^{tA}u(0).$$

Much depends on whether $[A^\dagger, A] = 0$. Matrices that satisfy this condition are said to be *normal*.

For such matrices we can combine the two exponentials in the middle to get

$$|u(t)|^2 = u^\dagger(0)e^{2tA_1}u(0).$$

Here $A_1 = \frac{A^\dagger + A}{2}$ is a hermitean matrix. Its eigenvalues are real. As long they are all less than 0,

$$|u(t)| \leq |u(0)|.$$

8.5   Normal operators

Let us understand this a bit more explicitly. Any matrix can be written as a linear combination of hermitian matrices

$$A = A_1 + iA_2.$$

Just put

$$A_1 = \frac{A^\dagger + A}{2}, \quad A_2 = \frac{A - A^\dagger}{2i}.$$

If the matrix is normal,

$$[A_1, A_2] = 0.$$

Then we can diagonalize A_1, A_2 by the *same* unitary transformation. In other words there is a unitary transformation that diagonalizes A itself:

$$A = S^\dagger \Lambda S, \quad S^\dagger S = 1, \quad \Lambda = \begin{pmatrix} \lambda_1 & 0 & 0 \\ \cdot & \cdot & \cdot \\ 0 & 0 & \lambda_n \end{pmatrix}.$$

The eigenvalues λ_i are in general complex; the real parts are eigenvalues of A_1 and the imaginary parts those of A_2. Each eigenvalue λ_i corresponds to an eigenvector

$$A\psi_i = \lambda_i \psi_i.$$

Unequal eigenvalues correspond to orthogonal vectors

$$\lambda_i \neq \lambda_j \implies \psi_i^\dagger \psi_j = 0.$$

Even if there are degenerate eigenvalues we can still choose a basis of eigenvectors that is orthonormal:

$$\psi_i^\dagger \psi_j = \delta_{ij}.$$

In this case we can expand

$$u(0) = \sum_i u_i(0)\psi_i,$$

$$u(t) = \sum_i e^{t\lambda_i} u_i(0)\psi_i,$$

$$|u(t)|^2 = \sum_i |u_i(0)|^2 e^{2t\operatorname{Re}\lambda_i}.$$

Thus, as long as the eigenvalues all have a non-negative real part,

$$\operatorname{Re}\lambda_i \leq 0,$$

we have solutions that do not grow in time:

$$|u(t)|^2 \leq |u(0)|^2.$$

The most familiar examples of normal operators are hermitean or anti-hermitean operators; then either A_1 or A_2 is zero. When $A_1 = 0$, we have the additional consequence that the eigenvalues are all purely imaginary. This correspond to solutions that are oscillatory.

Example 8.1

An example of a normal matrix is

$$A = \begin{pmatrix} -1 & i \\ i & -1 \end{pmatrix}$$

It has eigenvalues $-1 \pm i$. Both have negative real part. In this case the equation $\frac{du}{dt} = Au$ has bounded solutions $|u(t)| \leq |u(0)|$.

Exercise 8.3 Show that $[A^\dagger, A] = 0$ when $A = \begin{pmatrix} -a & b \\ b & -a \end{pmatrix}$ for complex numbers a, b.

What are the eigenvalues? What are the conditions on a, b in order that the linear equation $\frac{du}{dt} = Au$ have solutions satisfying $|u(t)| \leq |u(0)|$ for any initial condition?

8.6   A non-normal operator

An example of a linear ODE with an operator that is not normal is (Trefethen et al., 1993)

$$\frac{du}{dt} = Au, \quad u(0) = \begin{pmatrix} 1 \\ 0 \end{pmatrix}, \quad A = \begin{pmatrix} -a & 0 \\ 1 & -2a \end{pmatrix}, \quad a > 0. \tag{8.1}$$

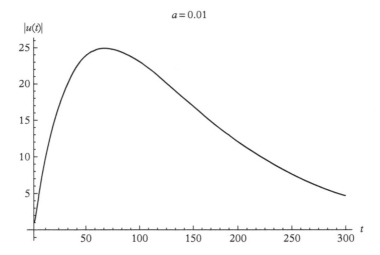

Figure 8.1 *Solution to a linear equation with a non-normal operator (eqn 8.1) grows at first before dying out: a transient*

Its eigenvalues are $-a, -2a$: both lie in the left half-plane (i.e., the stable half). But we can see that the solutions to the equation grow for a while before dying out.

The solution to eqn (8.1) is easy enough to find:

$$u(t) = \begin{pmatrix} e^{-at} \\ -\frac{1}{a}\left[e^{-2at} - e^{-at}\right] \end{pmatrix} \implies |u(t)|^2 = e^{-2at} + \frac{1}{a^2}[e^{-at} - e^{-2at}]^2$$

This falls off to zero as $t \to \infty$. But it grows to a norm of order $\frac{1}{a}$ (at a time of order $\frac{1}{a}$) before dying out (see Figure 8.1). This is an example of a transient. For small a the solution becomes large enough to invalidate the linear approximation.

The point of this exercise is that the linearization of the NS equations is a non-normal operator (Orr–Sommerfeld). So it has growing transients. Thus, even if there is no unstable spectrum, the perturbations can grow large enough to trigger nonlinear effects. This mechanism is believed to explain the instability of pipe Poiseuille flow.

8.7   A nonlinear model with transients

A simple model for the NS equations studied by (Trefethen et al., 1993) is

$$\frac{du}{dt} = Au + |u|Bu, \quad u(0) = \begin{pmatrix} \epsilon \\ 0 \end{pmatrix}, \tag{8.2}$$

where

$$B = \begin{pmatrix} 0 & 1 \\ -1 & 0 \end{pmatrix}, \quad A = \begin{pmatrix} -a & 0 \\ 1 & -2a \end{pmatrix}, \quad a > 0.$$

Of course, the NS equations are vastly more complicated, with an infinite number of unknowns. But studying such simple examples is an important part of theoretical physics. The nonlinear term is chosen such that by itself it will not change $|u|$. Because $B^\dagger = -B$:

$$\frac{du}{dt} = |u|Bu \implies \frac{d|u|}{dt} = 0.$$

This mimics the NS equations: the nonlinear tern $v \cdot \nabla v$ preserves energy; the dissipative term $\nu \nabla^2 v$ is linear.

If the initial condition has small enough norm $|u(0)|$, the solutions ought to die out for large time: the nonlinearity has only a small effect. But as $|u(0)|$ grows, the transients can get large enough that nonlinearity is important.

....

Exercise 8.4 Solve the above equation (8.2) numerically and plot the solution for various values of ϵ, a.

....

Numerical solution shows that for small ϵ the solutions of eqn (8.2) do die out. For ϵ larger than a critical value $> \epsilon_c$ (which depends on a), it grows to a value of order 1 for large time. See Figure 8.2.

The critical solution has $\epsilon \equiv u_1(0) \sim a^3, u_2(0) = 0$. The transient grows u_2 so that the solution becomes

$$u_1 \approx 2a^3, \quad u_2 \approx 2a^2$$

at intermediate values of t. It stays there for a while while the nonlinearity amplifies u_1 and the solution tends to

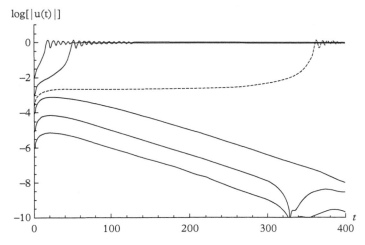

Figure 8.2 *Transients: The logarithm of the norm of u for $\epsilon = 10^{-6}, 10^{-5}, \cdots 10^{-2}$ and $a = \frac{1}{30}$ are the solid curves. The dotted curve is near the border between these two behaviors, for $\epsilon \approx 2.445 \times 10^{-4}$*

$$u_1 \approx 1, \quad u_2 \approx a$$

asymptotically, after some oscillations.

We see here a possible mechanism for a small perturbation to grow: a transient grows it to a value large enough that the nonlinearity kicks in, which then boosts it to an asymptotically constant solution. Analytical and numerical studies seem to confirm this picture (Aasen and Kreiss, 2006) of what happens in pipe Poiseuille flow.

....

Exercise 8.5 Find the asymptotic constant vector to which the solutions of the above equation tend as $t \to \infty$ and $\epsilon > \epsilon_c$.

Solution You should get $|u| = \frac{1}{2}\left(1 + \sqrt{1 + 8a^2}\right)$ and $\frac{u_2}{u_1} = \frac{a}{|u|}$.

....

8.8   Stability regained

In the mechanics of non-dissipative systems, an equilibrium point is an extremum of the potential. It is stable if the extremum is a minimum; and unstable if it is a maximum. The most basic example is a pendulum.

A simple pendulum has two equilibrium points: the bottom one which is stable. And another at the top, where it is unstable. There is a similar situation in fluid mechanics when two fluids of different densities coexist. If the heavier fluid is at the bottom and the lighter one at the top, this is a stable situation. If the heavier fluid is at the top it is unstable. It will seep into the lighter fluid through "fingers" until the situation is reversed. This is the Rayleigh–Taylor instability instability of Section 8.1.

To understand this better, let us reconsider the simple pendulum. The Lagrangian is

$$L = \frac{1}{2}ml^2\dot{\theta}^2 - mgl[1 - \cos\theta].$$

This can be written as

$$L = ml^2\left\{\frac{1}{2}\dot{\theta}^2 - \omega^2\left[1 - \cos\theta\right]\right\}, \quad \omega = \sqrt{\frac{g}{l}}.$$

It will be convenient to drop the overall factor of ml^2 as it has no effect on the equations of motion:

$$L_0 = \frac{1}{2}\dot{\theta}^2 - \omega^2\left[1 - \cos\theta\right].$$

We get the equation of motion

$$\ddot{\theta} + \omega^2 \sin\theta = 0.$$

There is a minimum for the potential at $\theta = 0$ and a maximum at $\theta = \pi$. The first is a stable equilibrium and the latter is unstable.

If $|\theta| << 1$ the equations become linear,

$$\ddot{\theta} + \omega^2 \theta = 0,$$

with solutions that are periodic:

$$\theta = A \cos\omega[t - t_0].$$

Near the other equilibrium point, we set $\chi = \theta - \pi$ to be small:

$$\ddot{\chi} - \omega^2 \chi = 0$$

which has exponentially growing solutions instead. Of course, χ can't really grow exponentially as it is a bounded variable. The linear approximation will break down as soon as χ is of order 1.

Suppose you had to find a way to stabilize the pendulum at its inverted position. How would you do it? When you stand up you are an inverted pendulum. You are able to stand stably only because of constant small adjustments made by your brain, based on measurements of attitude made in the inner ear. The first thing that happens when you lose consciousness is that you fall down. Active control through a feedback loop is one way to make an inverted pendulum stable.

But there is another way. If you were to vibrate the point of suspension at a frequency large compared to that of the pendulum, the effective potential energy is the average over these vibrations. This average can have a minimum at the top, leading to a stable inverted position. This curious gadget is called the Kapitza pendulum. To understand it we develop a theory of response of a system to a rapidly oscillating external force, simplifying section 30 of the classic work by (Landau and Lifshitz, 1977).

8.9 Rapidly changing external force

Consider a system with Lagrangian

$$L_0 = \frac{1}{2} g_{ij} \dot{q}^i \dot{q}^j - V(q).$$

We assume that g_{ij} is independent of q for simplicity (good enough for the pendulum). There is not much extra work in considering a system with many degrees of freedom.

Now suppose there is a correction to this,

$$L = L_0 - V_1(q, t),$$

where V_1 is a rapidly varying function of time which averages to zero over the characteristic time of the original Lagrangian (e.g., the period of an orbit):

$$<V_1(q, t)> = 0.$$

We do not assume that $V_1(q, t)$ is small or that it is periodic in time: only that it varies rapidly and has average zero. Even when V_1 is not small, it can have a small effect because it changes sign so often, canceling out its own effect. We can then suppose that the path q is a sum of a small rapidly varying part ξ of zero average, and a slowly varying function Q:

$$q^i(t) = Q^i(t) + \xi^i(t),$$

$$L = \frac{1}{2} g_{ij} \dot{Q}^i \dot{Q}^j - V(Q) + g_{ij} \dot{Q}^i \dot{\xi}^j - \xi^i \partial_i V(Q) - V_1(Q, t)$$

$$+ \frac{1}{2} g_{ij} \dot{\xi}^i \dot{\xi}^j - \xi^i \partial_i V_1(Q, t) \cdots.$$

We can now derive an equation for ξ, solve it, substitute it into L and average over a characteristic time of L_0 to get an effective Lagrangian $\bar{L}$. Terms containing a single ξ or $\dot{\xi}$ average to zero:

$$g_{ij} \ddot{\xi}^j = -\partial_i V_1(Q, t),$$

$$\bar{L} = \frac{1}{2} g_{ij} \dot{Q}^i \dot{Q}^j - V(Q) + \frac{1}{2} g_{ij} < \dot{\xi}^i \dot{\xi}^j > - <\xi^i \partial_i V_1(Q, t)>.$$

Now,

$$\xi^i \partial_i V_1(Q, t) = -g_{ij} \xi^i \ddot{\xi}^j$$

$$= g_{ij} \dot{\xi}^i \dot{\xi}^j - \frac{d}{dt} \left[g_{ij} \xi^i \dot{\xi}^j \right].$$

The time average over a long time period of a total derivative is zero. So

$$<\xi^i \partial_i V_1(Q, t)> = <g_{ij} \dot{\xi}^i \dot{\xi}^j>$$

Thus

$$\bar{L} = \frac{1}{2} g_{ij} \dot{Q}^i \dot{Q}^j - \bar{V}(Q),$$

where the effective potential has a correction equal to the average kinetic energy of the rapid motion:

$$\bar{V}(Q) = V(Q) + \frac{1}{2}g_{ij}<\dot{\xi}^i\dot{\xi}^j>.$$

In solving for ξ we can treat Q as a constant, as it varies so slowly:

$$\dot{\xi}^i \approx -g^{ij}\partial_j W_1(Q,t), \quad \frac{\partial W_1}{\partial t} = V_1,$$

giving us the effective potential

$$\bar{V} = V + \frac{1}{2}g^{ij}<\partial_i W_1 \partial_j W_1>.$$

8.10   The Kapitza pendulum

Let us now apply the above formalism to a pendulum whose point of support $(A(t), B(t))$ is oscillating rapidly.

The position of the pendulum bob is $(A + l\sin\theta, B + l\cos\theta)$. The angle θ is measured from the vertical line passing through the point of support. Then the kinetic energy is

$$K = \frac{1}{2}m\left[(\dot{A} - l\sin\theta\dot{\theta})^2 + (\dot{B} + l\sin\theta\dot{\theta})^2\right]$$

$$= \frac{1}{2}ml^2\dot{\theta}^2 - ml\left[\ddot{A}\sin\theta + \ddot{B}\cos\theta\right] + \cdots,$$

the dots representing total derivatives as well as terms independent of θ (which don't matter to the equation of motion for θ). The potential energy is, as before, $mgl[1 - \cos\theta]$. Again we factor out an overall factor of ml^2, leading to the Lagrangian

$$L = L_0 - V_1,$$

$$L_0 = \frac{1}{2}\dot{\theta}^2 - \frac{g}{l}[1 - \cos\theta], \quad V_1 = \frac{1}{l}\left[\ddot{A}\sin\theta + \ddot{B}\cos\theta\right].$$

We can split the motion of the pendulum $\theta(t) = \Theta(t) + \xi(t)$ into a slowly varying $\Theta(t)$ and a rapidly varying $\xi(t)$. Treating Θ as approximately a constant over a period of A, B, we get

$$W_1 = \frac{1}{l}\left[\dot{A}\sin\Theta + \dot{B}\cos\Theta\right]$$

and

$$\bar{V}(\Theta) = \frac{g}{l}\left[1 - \cos\Theta\right] + \frac{1}{2l^2}\left[<\dot{A}^2> \cos^2\Theta + <\dot{B}^2> \sin^2\Theta - 2 <\dot{A}\dot{B}> \sin\Theta\cos\Theta\right].$$

Dropping a term independent of Θ,

$$\bar{V}(\Theta) = \frac{g}{l}\left[1 - \cos\Theta\right] + \frac{1}{4l^2}\left[\left(<\dot{A}^2> - <\dot{B}^2>\right)\cos 2\Theta - 2 <\dot{A}\dot{B}> \sin 2\Theta\right].$$

The unperturbed pendulum has a stable equilibrium at $\Theta = 0$ and an unstable equilibrium at $\Theta = \pi$. These remain as equilibrium points in the case of purely vertical or horizontal motion ($\dot{A}\dot{B} = 0$).

There are in addition two more extrema if the support oscillates fast enough, at points satisfying

$$\cos\Theta = \left\{ \begin{array}{ll} \frac{gl}{\dot{A}^2} & \text{horizontal} \\ -\frac{gl}{\dot{B}^2} & \text{vertical} \end{array} \right\}.$$

- For rapid horizontal motion of the support, with $B = 0, \dot{A}^2 > gl$ the extremum at $\Theta = 0$ is unstable: the two new extrema to each side are stable. The pendulum will (with friction) eventually comes to rest at one these points instead of hanging down as usual.

- For rapid enough vertical motion of the support $A = 0, \dot{B}^2 > gl$, the extremum at the top becomes stable: $\bar{V}''(\pi) = -\frac{g}{l} + \frac{\dot{B}^2}{l^2} > 0$. The two new extrema on either side are unstable. This allows a pendulum with vertically oscillating support to stand on its head, and even oscillate around the point $\Theta = \pi$.

Thus a fast perturbation can turn an unstable equilibrium point into a stable one or vice versa. Remarkably, there is way to use this idea to counter the Rayleigh–Taylor instability instability (Wolf, 1969). Can this be of use in inertial confinement fusion where this instability is a nuisance?

....

Exercise 8.6 Research project: Kapitza pendulum Write a program that numerically solves the equations for a pendulum with oscillating support. Display the solution using an animation. Investigate the regions of stability and instability.

....

9

Integrable Models

Much of theoretical physics is about finding simple, solvable, models that illustrate some phenomenon or concept. The most important of these (the hydrogen atom, the simple harmonic oscillator) are also good approximations to common physical systems.

There are also models which are not directly relevant to nature (2D Ising model, the Abelian Higgs model, Bethe ansatz for spin chains) which form the basis of our understanding of important phenomena (phase transitions, massive gauge fields, etc.). As long as you keep in mind their limitations, they can be extremely useful tools. We will study a couple of such examples in this chapter.

9.1 KdV

A nonlinear equation that describes water waves in shallow channels was derived by Korteweg and de Vries (KdV) for the height ϕ of the wave:

$$\frac{\partial \phi}{\partial t} + \frac{\partial^3 \phi}{\partial x^3} + 6\phi \frac{\partial \phi}{\partial x} = 0.$$

This equation is written in a dimensionless form. There is no great significance to the value of the constant being 6 in the last term: by a rescaling of ϕ we can choose it to be any non-zero number. The choice here is convenient in calculations below.

If we ignore the nonlinear term, this describes linear waves $\phi(x, t) = e^{i[kx+\omega t]}$ with the dispersion relation

$$\omega = k^3.$$

Thus, the group velocity is $v_g = \frac{d\omega}{dk} = 3k^2$: waves of long wavelength (small k) move slowly. When we include the nonlinear effects, the speed of the wave can depend also on its amplitude.

Fluid Mechanics: A Geometrical Point of View. S. G. Rajeev © S. G. Rajeev 2018.
Published in 2018 by Oxford University Press. DOI: 10.1093/oso/9780198805021.001.0001

9.2   The soliton solution

Let us make the ansatz

$$\phi(x,t) = f(x - ut).$$

The equation gives

$$-uf' + f''' + 6ff' = 0 \implies$$
$$-uf + f'' + 3f^2 = c_1.$$

Multiplying by ϕ' and integrating again,

$$-\frac{1}{2}uf^2 + \frac{1}{2}f'^2 + f^3 = c_2 + c_1 f.$$

Let us impose the b.c. that ϕ, ϕ' tend to zero at infinity. Then $c_1 = c_2 = 0$ and

$$-\frac{1}{2}uf^2 + \frac{1}{2}f'^2 + f^3 = 0,$$

which is easily integrated by a substitution involving hyperbolic functions. Or it can be solved by the ansatz

$$\phi(x) = A \cosh^{-2} Bx.$$

Substitution into the equation gives

$$A = \frac{u}{2}, \quad B = \frac{\sqrt{u}}{2},$$
$$\phi(x,t) = \frac{u}{2}\operatorname{sech}^2 \frac{\sqrt{u}}{2}(x - ut).$$

Thus, the amplitude of the wave is proportional to its speed. This is an example of a soliton: a nonlinear wave whose shape is determined by dispersion and focusing (nonlinearity). A more remarkable fact is that we can find an exact solution representing the collision of two such solitons, which then emerge with unchanged shapes. This is most unusual as most nonlinear equations would lead to collisions that would dissipate through emitting small amplitude waves ("radiation").

....

Exercise 9.1 Find a solution $f(x)$ that is periodic in x to the above ODE, in terms of elliptic functions. This would describe a periodic wave of solitonic pulses.

....

9.3 Multi-soliton solutions

The systematic way to generate solutions of KdV uses the "inverse scattering" method we will briefly describe later. We just note an outcome of this more elaborate theory, a way to nonlinearly superpose soliton solutions.

The one-soliton solution can be written also as

$$\phi_1(x, t) = 2 \frac{\partial^2}{\partial x^2} \log K_1(x, t), \quad K_1(x, t) = 1 + \frac{1}{2a} e^{2a[x - 4a^2 t + \delta]}.$$

The constant a is related to velocity $u = 4a^2$. Also, δ parametrizes the initial position of the soliton. Now suppose we define an $n \times n$ matrix depending on $2n$ parameters a_i, δ_i for $i = 1, \cdots n$:

$$B_{ij} = \delta_{ij} + \frac{1}{a_i + a_j} e^{\xi_i + \xi_j}, \quad \xi_i = 2a_i[x - 4a_i^2 t + \delta_i].$$

Then, it turns out that

$$\phi_n(x, t) = 2 \frac{\partial^2}{\partial x^2} \log K_n(x, t), \quad K_n(x, t) = \det B$$

is a solution to the KdV equation! Again, the parameters a_i give the velocity of the ith soliton and δ_i parametrizes its position.

The purpose of this summary is to entice you into studying the beautiful mathematical theory of integrable PDEs and solitons further. There are any number of books on this subject (e.g., Whitham, 1999).

9.4 Lax pair

Suppose we define two linear differential operators

$$L\psi = \psi'' + \phi\psi,$$
$$P\psi = 4\psi''' + 6\phi\psi' + 3\phi'\psi.$$

By coincidence, L is the Schrödinger operator of quantum mechanics. P is a new beast.

....

Exercise 9.2 Show that

$$PL\psi - LP\psi = \left[\phi''' + 6\phi\phi'\right]\psi.$$

....

The operators P, L are chosen so that their commutator is the spatial part of the KdV equation. That is, the KdV equation can be written as an equation for the evolution of L as a function of t. (This has nothing to do with time evolution in quantum mechanics.)

$$\frac{\partial L}{\partial t} + [P, L] = 0 \iff \frac{\partial \phi}{\partial t} + \frac{\partial^3 \phi}{\partial x^3} + 6\phi \frac{\partial \phi}{\partial x} = 0.$$

Another, equivalent, point of view: the KdV equation is a condition that the pair of equations

$$\frac{\partial^2 \psi_\lambda(x, t)}{\partial x^2} + \phi(x, t)\psi_\lambda(x, t) = \lambda \psi_\lambda(x, t),$$

$$\frac{\partial \psi_\lambda(x, t)}{\partial t} = 4\frac{\partial^3 \psi_\lambda(x, t)}{\partial x^3} + 6\phi(x, t)\frac{\partial \psi_\lambda(x, t)}{\partial x} + 3\frac{\partial \phi(x, t)}{\partial x}\psi_\lambda(x, t)$$

have a simultaneous solution. Here λ is an auxiliary variable: it is independent of t.

Now, suppose $\phi(x) \to 0$ as $|x| \to \infty$. Then there is a solution of the first equation such that $\psi_\lambda(x, t) \to e^{\sqrt{\lambda}x} + R(\lambda, t)e^{-\sqrt{\lambda}x}$ as $x \to \infty$ and $\psi(x) \to T(\lambda, t)e^{\sqrt{\lambda}x}$ as $x \to \infty$. (λ is a complex variable and we choose the principal branch of the square root.) The second equation now implies that the "scattering data" depend in a simple way on time:

$$R(\lambda, t) = R(\lambda, 0)e^{-4\lambda^{\frac{3}{2}}t},$$

$$T(\lambda, t) = T(\lambda, 0)e^{4\lambda^{\frac{3}{2}}t}.$$

The initial "potential" $\phi(x, 0)$ determines $R(\lambda, 0)$ and $T(\lambda, 0)$. Remarkably, it is possible to go the other way: reconstruct the potential from the the scattering data. (It involves solving a system of *linear* integral equations.)

So we can solve reduce the solution of KdV to a sequence of linear problems:

1. Solve the Schrödinger scaterring problem for the initial data $\phi(x, 0)$ to find $R(\lambda, 0)$ and $T(\lambda, 0)$.

2. Solve the inverse scattering problem to determine $\phi(x, t)$ from the data $R(\lambda, 0)e^{4\lambda^{\frac{3}{2}}t}$ and $T(\lambda, 0)e^{4\lambda^{\frac{3}{2}}t}$.

Thus determining the scattering data is an analog of the Fourier transformation for linear PDEs. Bound states of the Schrödinger equation correspond to poles in the scattering amplitude; the soliton solutions of KdV arise from them. For more see (Whitham, 1999).

These methods are very special to the KdV and a handful of other PDEs. Typical equations of fluid mechanics are far from integrable. But they help us understand how waves of large amplitude propagate.

9.5   Hamiltonian formalism of Fadeev and Zakharov

The KdV equation can be thought of as an infinite dimensional mechanical system, with a hamiltonian and a Poisson bracket. This is true of more realistic ideal fluid models as well, but they are more complicated. We will return to this issue later. It is always a good idea to work out a simple case first.

Write the KdV equation in the form of a conservation law:

$$\frac{\partial \phi}{\partial t} + \frac{\partial}{\partial x}\left[\phi'' + 3\phi^2\right] = 0.$$

Thus, the quantity $\int \phi \, dx$ is independent of time. Moreover, if we set

$$H = \int \left[\frac{1}{2}\phi'^2 - \phi^3\right] dx$$

we have

$$\frac{\delta H}{\delta \phi(x)} = -\phi'' - 3\phi^2,$$

so the KdV equation can be written as

$$\frac{\partial \phi}{\partial t} = \frac{\partial}{\partial x}\frac{\delta H}{\delta \phi(x)}.$$

This has an interesting meaning. If we think of H as the hamiltonian, the equations of motion are of the form

$$\frac{\partial \phi}{\partial t} = \{H, \phi(x)\},$$

with the Poisson brackets

$$\{F(\phi), \phi(x)\} = \frac{\partial}{\partial x}\frac{\delta F}{\delta \phi(x)}$$

for any function $F(\phi)$ of ϕ. An equivalent statement is

$$\{\phi(y), \phi(x)\} = \frac{\partial}{\partial x}\delta(x - y).$$

As a bonus we get the fact that H is conserved (which we could have checked directly using the equations of motion).

KdV equations are beloved by mathematical physicists because they are a rare example of a nonlinear *integrable* dynamical system with a large number of degrees of freedom:

there is a canonical transformation, to variables (the scattering data we encountered earlier) that have a simple time evolution. This transformation is thus an analog of the normal mode decomposition of a system of coupled harmonic oscillators, rarely possible in nonlinear systems.

9.6   The hamiltonian formalism of Magri

There is a different choice of Poisson bracket and hamiltonian that also gives the KdV equation.

$$H_2 = \frac{1}{2} \int \phi^2 \, dx,$$

$$\{F(\phi), \phi(x)\} = -\frac{\partial^3}{\partial x^3} \frac{\delta F}{\delta \phi(x)} - 2\phi(x) \frac{\partial}{\partial x} \frac{\delta F}{\delta \phi(x)} - \frac{\delta F}{\delta \phi(x)} \frac{\partial}{\partial x} \phi.$$

It is not obvious that this Poisson bracket satisfies the Jacobi identity. It can be verified by a tedious calculation. Or by a slick method that shows that this is the Kirillov–Poisson bracket on the co-adjoint orbit of the Virasoro algebra. (More on Poisson brackets and Lie algebras later).

It is rare that the same dynamical system has two different hamiltonian formalisms. They are called bi-hamiltonian systems. By going back and forth between the two we can construct an infinite number of conservation laws. This explains the integrability of KdV.

....

Exercise 9.3 Show by direct computation that $\frac{1}{2} \int \phi^2 \, dx$ is conserved for the KdV equation.

   Solution Multiply the equation by ϕ. Express the terms involving spatial derivatives as a total derivative. Integrate over x.

....

9.7   The vortex filament

We have all seen smoke rings move through air without much change of shape. Although eventually they get deformed and dissolve away, for a while vortices can evolve coherently. The simplest case is a very long vortex with small curvature. It can be maintained by some input of energy to compensate for the loss due to dissipation at the core, where the velocity gradients are large. An approximate equation can be derived for the evolution of such a vortex, with a beautiful geometric meaning. To appreciate it, we digress to introduce some differential geometry of curves.

9.8   Geometry of curves

The main reference for this beautiful subject is the book by (Do Carmo, 1976). A *curve* is a smooth function $\xi : [a, b] \to \mathbb{R}^3$. That is, at each value of some parameter σ, we have a point in space $\xi(\sigma)$. The parameter itself need not have any geometric meaning: a change $\sigma \to \sigma' = \phi(\sigma)$ will give an equally good parametrization as long as ϕ is smooth with a smooth inverse.

Among all curves connecting two points, the straight line has the shortest length. We can state this as a variational principle for the length

$$L = \int \sqrt{\frac{d\xi}{d\sigma} \cdot \frac{d\xi}{d\sigma}} \, d\sigma.$$

The condition for L to have a minimum is

$$\frac{d}{d\sigma} \left[\frac{1}{2\sqrt{\frac{d\xi}{d\sigma} \cdot \frac{d\xi}{d\sigma}}} \frac{d\xi}{d\sigma} \right] = 0.$$

This equation looks nicer if we make the change the parameter to the arc length:

$$s(\sigma) = \int_a^\sigma \sqrt{\frac{d\xi}{d\sigma} \cdot \frac{d\xi}{d\sigma}} \, d\sigma, \implies \frac{ds}{d\sigma} = \sqrt{\frac{d\xi}{d\sigma} \cdot \frac{dx\xi}{d\sigma}},$$

$$\frac{d\xi}{ds} = \frac{\frac{d\xi}{d\sigma}}{\frac{ds}{d\sigma}} = \frac{1}{\sqrt{\frac{d\xi}{d\sigma} \cdot \frac{d\xi}{d\sigma}}} \frac{d\xi}{d\sigma}.$$

The equation for a straight line becomes just

$$\frac{d^2\xi}{ds^2} = 0.$$

Thus the arc length is a natural choice of parameter. Then $\xi' \equiv \frac{d\xi}{ds}$ points along the tangent to the curve and has unit length:

$$\xi' \cdot \xi' = 1.$$

The magnitude $|\xi''| = \kappa$ measures how much the curve departs from being a straight line: the *curvature*. Differentiating the above equation, we see that the second derivative points normal to the tangent:

$$\xi'' \cdot \xi' = 0.$$

The vector $n = \frac{\xi''}{\kappa}$ is the unit normal. Differentiating $n \cdot n = 1, n \cdot \xi' = 0$ we get

$$n' \cdot n = 0, \quad n' \cdot \xi' + \kappa = 0.$$

If the curve were lying in some plane, n' would have to be a linear combination of n, ξ'. So we would have $n' = -\kappa \xi'$. For a space curve there could be a component of n' normal to both n and ξ'.

The cross poduct $\xi' \times n = b$ is the unit vector normal to both the tangent and to n; it is called the *binormal*. The three unit vectors (ξ', n, b) form a mutually orthogonal system. So we can expand

$$n' = -\kappa \xi' + \tau b.$$

The quantity τ is the *torsion*; it measures how much the curve twists away from the plane spanned by ξ', n. If the torsion is zero everywhere, the curve lies in some plane.[1]

We can combine the above notions into a system that describes the evolution of the triad along the curve:

$$\frac{d}{ds} \begin{pmatrix} \xi' \\ n \\ b \end{pmatrix} = \begin{pmatrix} 0 & \kappa & 0 \\ -\kappa & 0 & \tau \\ 0 & -\tau & 0 \end{pmatrix} \begin{pmatrix} \xi' \\ n \\ b \end{pmatrix}. \quad \text{Frenet–Serret equations}$$

The point of these equations is that the curvature and torsion determine the curve, up to some initial conditions. Thus the pair of functions $\kappa(s), \tau(s)$ give an "intrinsic" description of a curve, independent of the choice of coordinates.

We now want to apply this geometry to describe a vortex filament: a curve along which there is a concentration of vorticity.

....

Exercise 9.4 What is a curve of constant curvature and zero torsion? What if the curvature and torsion are both constant and non-zero?

....

9.9   Velocity from vorticity: the Biot–Savart law

Recall that for an incompressible fluid, vorticity determines velocity:

$$\text{curl } v = \omega, \quad \text{div } v = 0.$$

---

[1] The word torsion is used in a different sense in differential geometry: the commutator of covariant derivatives of a scalar. This is unrelated to the torsion of curves we use here.

In the case of a fluid that fills all of space (with the boundary condition that velocity and vorticity vanish at infinity) we can write an explicit integral formula for this relation. The problem is identical to determining the magnetic field of a stationary current density,

$$v(x) = \frac{1}{4\pi} \int \frac{(x-y) \times \omega(y)}{|x-y|^3} dy. \quad \text{Biot–Savart law}$$

The proof begins with the Green's function of the Laplacian

$$\nabla^2 \left[-\frac{1}{4\pi|x-y|} \right] = \delta(x-y).$$

Note the identity

$$\text{curl } \omega = -\nabla^2 v$$

so that

$$v(x) = \int \frac{1}{4\pi|x-y|} \text{curl } \omega(y) dy.$$

An integration by parts (use $\nabla \times (G\omega) = \nabla G \times \omega + G\nabla \times \omega$ and the b.c. at infinity) gives

$$v(x) = \int \nabla_y \left[\frac{1}{4\pi|x-y|} \right] \times \omega(y) dy$$

$$= -\int \nabla_x \left[\frac{1}{4\pi|x-y|} \right] \times \omega(y) dy$$

$$= \int \frac{(x-y)}{4\pi|x-y|^3} \times \omega(y) dy,$$

as we claimed.

9.10   The velocity field of a vortex filament

Suppose the vorticity is zero outside a thin region surrounding a curve: a vortex filament. The curve can change shape with time, so for each time t we have some curve $\xi(s,t)$ given as a function of its arc length s. The fluid circulates around this curve. The vorticity distribution is

$$\omega(x,t) = \Gamma \xi'(s,t)\delta(x-\xi(s,t)).$$

That is, the vorticity is concentrated on the curve $\xi(s, t)$ and points along its tangent. By conservation of vorticity, the strength Γ is independent of time.

The velocity field away from the filament is determined by the Biot–Savart law (it is much like the magnetic field created by a thin wire carrying a current Γ):

$$v(x, t) = \frac{\Gamma}{4\pi} \int \frac{[x - \xi(s, t)]}{|x - \xi(s, t)|^3} \times \xi'(s, t) ds.$$

If we have several such filaments $\xi_k(s, t)$ we just have to add the velocity field created by each to get the velocity field away from all of them:

$$v(x, t) = \sum_k \frac{\Gamma_k}{4\pi} \int \frac{[x - \xi_k(s, t)]}{|x - \xi_k(s, t)|^3} \times \xi'_k(s, t) ds.$$

So far so good. Our agenda here is to use this formula to derive some equations similar to the Kirchoff equations for two-dimensional vortex dynamics: eliminate velocity entirely and describe the fluid as a collection of vortex filaments. For this we have to say that the time derivative of the position of the filament is equal to the velocity field at its position:

$$\frac{\partial \xi_k(s, t)}{\partial t} = v(\xi_k(s), t).$$

But we run into a problem: our formula for velocity is not valid at the filaments themselves. For example, in the case of a single filament, if we try to calculate the velocity at some point on the filament we get an integral that diverges:

$$v(\xi(s), t) = \frac{\Gamma}{4\pi} \int \frac{[\xi(s) - \xi(s_1, t)]}{|\xi(s) - \xi(s_1, t)|^3} \times \xi'(s_1, t) ds_1.$$

The divergence comes from the region where $s_1 \to s$ because the denominator vanishes.

9.11   Regularization and renormalization

The point is that no vortex filament can be infinitesimally thin: it has a core where the vorticity is finite but large. The delta function above should be replaced by some positive function that has integral one and with a maximum at the origin:

$$\delta_a(x) > 0, \quad \int \delta_a(x) dx = 1, \quad \lim_{a \to 0} \delta_a(x) = \delta(x).$$

We must choose this core model such that we get a manageable formula for velocity. One such regularization is to replace the integral above by

$$\int \frac{[\xi(s) - \xi(s_1, t)] \times \xi'(s_1, t)}{\left\{[\xi(s) - \xi(s_1, t)]^2 + a^2\right\}^{\frac{3}{2}}} ds_1,$$

where a measures the size of the core. Working back, you can verify that this amounts to the following model for the core (regularization of the delta function):

$$\delta_a(x) = \frac{3a^2}{4\pi \left(a^2 + x^2\right)^{5/2}}.$$

Split the integral into a part near the singularity and a part that remains finite:

$$v(\xi(s), t) = \frac{\Gamma(\epsilon)}{4\pi} \int_{|s-s_1|<\epsilon a} \frac{[\xi(s) - \xi(s_1, t)] \times \xi'(s_1, t)}{\left\{[\xi(s) - \xi(s_1, t)]^2 + a^2\right\}^{\frac{3}{2}}} ds_1 +$$

$$\frac{\Gamma(\epsilon)}{4\pi} \int_{|s-s_1|>\epsilon a} \frac{[\xi(s) - \xi(s_1, t)] \times \xi'(s_1, t)}{\left\{[\xi(s) - \xi(s_1, t)]^2 + a^2\right\}^{\frac{3}{2}}} ds_1.$$

Here ϵ is some small cutoff. We allow the strength of the vortex $\Gamma(\epsilon)$ to depend on its cutoff: it will be chosen so that the velocity field has a finite limit as $\epsilon \to 0$. (The jargon borrowed from quantum field theory is that we "renormalize" the vortex strength.) If we expand the integrand around $s_1 = s$ to second order in the Taylor series (and use the facts $\xi'^2 = 1$, $\xi' \cdot \xi'' = 0$):

$$\frac{[\xi(s) - \xi(s_1, t)] \times \xi'(s_1, t)}{\left\{[\xi(s) - \xi(s_1, t)]^2 + a^2\right\}^{\frac{3}{2}}} \approx$$

$$\frac{-\frac{1}{2}(s_1 - s)^2 \xi''(s) \times \xi'(s) - (s_1 - s)\xi'(s) \times (s_1 - s)\xi''(s)}{\left\{\left[(s_1 - s)\xi'(s) + \frac{1}{2}(s_1 - s)^2 \xi''(s)\right]^2 + a^2\right\}^{\frac{3}{2}}}$$

$$\approx \xi'(s) \times \xi''(s) \frac{-\frac{1}{2}(s - s_1)^2}{\left\{[s - s_1]^2 + a^2\right\}^{\frac{3}{2}}}.$$

Since

$$\int_{|s-s_1|<\epsilon a} \frac{(s - s_1)^2 ds_1}{\left\{[s - s_1]^2 + a^2\right\}^{\frac{3}{2}}} = \int_{-\epsilon}^{\epsilon} \frac{\sigma^2 d\sigma}{\left\{\sigma^2 + 1\right\}^{\frac{3}{2}}} \approx \log \epsilon^2 + \mathrm{O}\left(\epsilon^2\right)$$

we get

$$v(\xi(s), t) = \frac{\Gamma(\epsilon) \log \epsilon^2}{4\pi} \xi'(s) \times \xi''(s) + \frac{\Gamma(\epsilon)}{4\pi} \int_{|s-s_1|>\epsilon a} \frac{[\xi(s) - \xi(s_1, t)] \times \xi'(s_1, t)}{\{[\xi(s) - \xi(s_1, t)]^2 + a^2\}^{\frac{3}{2}}} ds_1.$$

Now we choose

$$\Gamma(\epsilon) = \frac{4\pi G}{\log \epsilon^2}$$

so that the coefficient of the first term has a finite limit. The constant G measures the strength of the vortex (this is the renormalization):

$$v(\xi(s), t) = G\xi'(s) \times \xi''(s) + \frac{G}{\log \epsilon^2} \int_{|s-s_1|>\epsilon a} \frac{[\xi(s) - \xi(s_1, t)] \times \xi'(s_1, t)}{\{[\xi(s) - \xi(s_1, t)]^2 + a^2\}^{\frac{3}{2}}} ds_1.$$

Thus as $\epsilon \to 0$, only the remnant of the divergent term survives:

$$v(\xi(s), t) = G\xi'(s) \times \xi''(s).$$

Setting this equal to the time derivative, we get an equation for the time evolution of the filament:

$$\frac{\partial \xi(s, t)}{\partial t} = G\xi' \times \xi''.$$

This equation has a beautiful meaning: the time derivative of the curve at any point is proportional to the curvature at that point, and is pointed in the direction of the binormal. Even more remarkably, (Hasimoto, 1972) has shown that this equation can be solved exactly. The key is the transformation

$$\psi(s, t) = \kappa(s, t) e^{i \int_0^s \tau(s_1, t) ds_1}.$$

This variable satisfies an equation that every good mathematical physicist knows how to solve by the inverse scattering method: the nonlinear Schrödinger equation:

$$-i\frac{\partial \psi}{\partial t} = \frac{\partial^2 \psi}{\partial s^2} + \frac{1}{2}|\psi|^2 \psi.$$

The nonlinear Schrödinger equation has a well-known soliton solution. One application of it is to describe pules of light transmitted along an optical fiber. These pulses carry most of the data across the fiber glass backbone of the internet. It is interesting to see what sort of vortex filament this solution represents (Hasimoto, 1972).

The torsion is some constant τ_0, meaning that the vortex winds around an axis at a constant rate. The solution depends on the combination $\eta(s, t) = s - 2\tau_0 t$, meaning that is a wave that propagates along the axis at a constant speed $2\tau_0$.

Figure 9.1 *The vortex filament soliton. The twist in the vortex moves the axis with time*

The curvature is

$$\kappa(s, t) = \frac{2v}{\cosh v\eta(s, t)}.$$

It vanishes for large η, so that the filament is a straight line at both ends. The parameter v measures the strength of the wave; the maximum curvature is $2v$. The solution is, explicitly,

$$\xi(s, t) = \begin{pmatrix} s - \frac{2v \tanh[v(s-2t\tau_0)]}{v^2+\tau_0^2} \\ \frac{2v \cos[s\tau_0+t(v^2-\tau_0^2)]\operatorname{sech}[v(s-2t\tau_0)]}{v^2+\tau_0^2} \\ \frac{2v \sin[s\tau_0+t(v^2-\tau_0^2)]\operatorname{sech}[v(s-2t\tau_0)]}{v^2+\tau_0^2} \end{pmatrix}.$$

For $v = 1.1, \tau_0 = 0.8, t = 1.8$ we plot it in Figure 9.1. As time evolves the twist in the filaments moves along the axis and the whole thing rotates around its axis.

....

Exercise 9.5 Write a program that plots the soliton vortex filament of (Hasimoto, 1972). Time dependence is best displayed by an animation.

....

9.12 Relation to the Heisenberg model

There is a surprising equivalence of the vortex equations to a model for magnetism due to Heisenberg. It has hamiltonian

$$H = \frac{1}{2} \int S'^2 dx.$$

$S(x)$ is a unit vector at each point on the line, representing the spin at that point. The hamiltonian represents the total cost in energy of nearby spins pointing different directions. The Poisson brackets of spin,

$$\{S_a(x), S_b(y)\} = -\epsilon_{abc} S_c(x) \delta(x - y),$$

lead to the equation of motion

$$\frac{\partial S_a}{\partial t} = \epsilon_{abc} S_b \frac{\delta H}{\delta S_c}$$

so that

$$S \cdot S = 1, \quad \frac{\partial S}{\partial t} = S \times S''. \quad \text{Heisenberg model}$$

If we differentiate the vortex filament equation,

$$\frac{\partial \xi'(s, t)}{\partial t} = G\xi' \times \xi'''.$$

If we identify $\xi' \equiv S$ and the arc length along the filament $s \equiv x$ with the variable that labels the position of the spin, we get the Heisenberg equation!

A state of the magnet with spins pointing in the same direction corresponds to a vortex filament that is simply a straight line. This is a static solution. Small vibrations of this straight filament correspond to spin waves. These satisfy the linear Schrödinger equation

$$S \approx \begin{pmatrix} \chi_1 \\ \chi_2 \\ 1 \end{pmatrix}, \quad \chi = \chi_1 + i\chi_2,$$

$$-i\frac{\partial \chi}{\partial t} = \chi''.$$

Some solutions are waves $\chi(x, t) = Ae^{i[kx - \omega t]}$ with the dispersion relation $\omega = k^2$.

There is a well-developed theory of how to solve the Heisenberg model which we can use to solve the vortex filament equation. We won't elaborate, only point to the authoritative treatment in Part 2, Chapter II, section 3 of the book by (Faddeev and Takhtajan, 1987).

10

Hamiltonian Systems Based on a Lie Algebra

The rigid body and the ideal fluid are very different systems physically. Yet there is a mathematical object that ties them together: a Lie algebra. Euler discovered both systems of equations, and must have sensed their unity. But it took another hundred years or so to bring this underlying principle to the surface. A guide to the literature is (Holm et al., 1998).

10.1 Rigid body mechanics

A rigid body is one which moves such that the distance between two particles in it remain fixed as it moves. It is a completely different thing from a fluid, where particles move about freely with no such constraint. We will just recall the basic facts as a more detailed discussion was given elsewhere (Rajeev, 2013).

A rigid body can move in two ways. Its center of mass can be translated; and it can rotate around some axis passing through this center of mass. Let us ignore the translational motion; it doesn't affect rotational motion and is easily taken into account later. The angular velocity Ω is a vector pointed along the axis of rotation, with a magnitude equal to the rate of change of the angle around it. The moment of inertia of the body is a symmetric tensor (3×3 matrix) G determined by the shape of the body:

$$G_{ij} = \frac{1}{2} \int \rho(x) \left[x^2 \delta_{ij} - x_i x_j \right] d^3 x.$$

There is a coordinate system in which this tensor is diagonal:

$$G = \begin{pmatrix} G_1 & 0 & 0 \\ 0 & G_2 & 0 \\ 0 & 0 & G_3 \end{pmatrix},$$

the eigenvalues G_i being the principal moments of inertia. In general they are unequal. It is important to remember that this coordinate system in which G is diagonal is not

Fluid Mechanics: A Geometrical Point of View. S. G. Rajeev © S. G. Rajeev 2018.
Published in 2018 by Oxford University Press. DOI: 10.1093/oso/9780198805021.001.0001

inertial: it rotates with the body. The time derivative of any vector X in the inertial frame is related to the time derivative in the co-rotating frame by

$$\left[\frac{dX}{dt}\right]_{\text{co-rotating}} + \Omega \times X = \left[\frac{dX}{dt}\right]_{\text{inertial}}.$$

This is analogous to the relation between the partial derivative and the material derivative in fluid mechanics.

The angular momentum is the product

$$L_i = G_{ij}\Omega_j.$$

Since G is invertible, there is a one–one relation between L and Ω. In the co-rotating frame, this simplifies to $L_1 = G_1\Omega_1$ etc.

In the absence of external torques, the angular momentum is conserved in an inertial frame. So we get Euler's equation for the rigid body in the co-rotating frame

$$\frac{dL}{dt} + \Omega \times L = 0, \quad \Omega_i = \frac{1}{G_i}L_i. \tag{10.1}$$

We can eliminate Ω to get

$$\frac{dL_1}{dt} + a_1 L_2 L_3 = 0, \quad a_1 = \frac{1}{G_2} - \frac{1}{G_3},$$

etc.

Now we see that the vorticity form of the Euler equations of an incompressible fluid is mathematically similar to the equations of a rigid body. The velocity potential of the fluid is analogous to Ω. The vorticity of the fluid is analogous to the angular momentum L of the rigid body. The Laplacian ∇^2 is like the moment of inertia tensor. Indeed, we can (by Fourier analysis) diagonalize the Laplacian and write the equations as a nonlinear system of ordinary differential equations (ODEs) as above. Of course the fluid is vastly more complicated, as there are an infinite number of unknown quantities.

This analogy goes even deeper: the Lie algebra of rotations of the rigid body corresponds to the Lie algebra under the commutator of vector fields; the hamiltonian of the fluid can be expressed in terms of vorticity, and so on. Arnold used these analogies to establish a deep theorem on the instability of fluids (Chapter 11). To understand these ideas we need to review some mathematical concepts.

10.2   Vector spaces

Recall that a vector space V is any set on which addition and multiplication by a real number ("scalar") is defined, and satisfy the usual properties

$$u + v = v + u, \quad \text{commutativity of addition}$$
$$u + (v + w) = (u + v) + w, \quad \text{associativity of addition}$$
$$\alpha(u + v) = \alpha u + \alpha v, \quad (\alpha + \beta)u = \alpha u + \beta u, \quad \alpha \in \mathbb{R}, \quad \text{distributivity}$$

The standard example are vectors in Euclidean space, where $v = (v_1, v_2, v_3)$ is a triple of scalars, the components along each coordinate axis. But the set of vector fields in some domain gives an example as well, as do functions on some domain.

10.2.1   Basis

Recall also that a set of linearly independent vectors form a basis if any vector can be written as linear combinations:

$$v = v^k e_k.$$

Under a change of basis the components transform contravariantly:

$$e'_k = S^l_k e_l,$$
$$v = v'^k e'_k = v^k e_k,$$
$$v'^k = v^l \left[S^{-1} \right]^k_l.$$

10.2.2   Metric

A metric (also called *scalar product* or *inner product*) is a function that takes pairs of vector to scalars such that

$$g(u, v) = g(v, u), \quad \text{symmetry}$$
$$g(u, \alpha v + \beta w) = \alpha g(u, v) + \beta g(u, w). \quad \text{linearity}$$

We will in addition assume positivity of the metric

$$g(u, u) \geq 0, \quad g(u, u) = 0 \iff u = 0.$$

The standard example is the scalar product $g(u, v) = \sum_i u_i v_i \equiv u \cdot v$ of vectors in Euclidean space. But any positive (i.e., symmetric with positive eigenvalues) matrix wil suffice

$$g(u, v) = g_{ij} u^i v^j.$$

Under changes of basis, the metric components transform covariantly:

$$g(u, v) = g_{ij} u^i v^j = g'_{ij} u'^i v'^j,$$
$$g'_{ij} = g_{kl} S^k_i S^l_j.$$

An important example is the moment of inertia of a rigid body, which need not have equal eigenvalues.

On the vector space of functions in some domain,

$$(f, g) = \int f(x)g(x)\,dx$$

will be an inner product. The main point of an inner product is that it allows one to talk of the length of a vector

$$|u| = \sqrt{< u, u >},$$

generalizing the familiar notion of length in Euclidean space. Now you see why we assumed positivity.

10.2.3   Metric components: covariant and contravariant

$$g_{kl} = g(e_k, e_l)$$

transform covariantly. The inverse matrix g^{kl} transforms contravariantly:

$$g'^{ij} = g^{kl} \left[S^{-1} \right]^i_k \left[S^{-1} \right]^j_l.$$

The metric tensor allows us to convert covariant components to a vector

$$v_k = g_{kl} v^l.$$

Thus if we identify the moment of inertia as the metric, the angular velocity represents the contravariant components and angular momentum the covariant components of some vector. We can also go the other way ("raise and index") using the inverse matrix

$$v^k = g^{kl} v_l.$$

10.3   Lie algebra

Just as the inner product generalizes the dot product, a Lie algebra generalizes the cross product.

A *Lie bracket* (also called *Lie product*) is a map $V \times V \to V$ satisfying

$$[u, \alpha v + \beta w] = \alpha[u, v] + \beta[u, w], \quad \text{linearity}$$
$$[u, v] = -[v, u], \quad \text{anti-symmetry}$$
$$[u, [v, w]] + [v, [w, u]] + [w, [u, v]] = 0. \quad \text{Jacobi identity}$$

We have already encountered several examples.

Example 10.1

The basic example is the cross product in $\mathbb{R}^3$. It is obvious that $u \times v = -v \times u$. The Jacobi identity is left as an exercise.

Example 10.2

Sophus Lie, a Norwegian mathematician, discovered this structure while studying the Poisson bracket (P.B.) of classical mechanics. The vector space of observables (functions on the phase space of a mechanical system) admits a bracket

$$\{f,g\} = \sum_{i=1}^{n} \left(\frac{\partial f}{\partial p_i} \frac{\partial g}{\partial q^i} - \frac{\partial f}{\partial q^i} \frac{\partial g}{\partial p_i} \right).$$

In particular we have the *canonical commutation relations*

$$\left\{ p_i, q^j \right\} = \delta_i^j, \quad \left\{ q^i, q^j \right\} = 0 = \left\{ p_i, p_j \right\}.$$

Most of the examples of interest to us arise as sub-algebras of the above example, that is, sets of functions whose P.B. are linear combinations of each other.

Example 10.3

The angular momentum components form a three-dimensional Lie algebra:

$$\{L_1, L_2\} = L_3, \quad \{L_2, L_3\} = L_1, \quad \{L_3, L_1\} = L_2.$$

For a single point particle $L = r \times p$ and the canonical P.B. give these relations for L. For the rigid body $L = \sum_a r_a \times p_a$ is the sum of angular momenta of each molecule. Since each terms satisfies the P.B. above and, different terms have zero P.B., we again get the result.

A moment's thought will show that this is in fact isomorphic (equivalent in structure) to our first example:

$$\{u \cdot L, v \cdot L\} = (u \times v) \cdot L$$

for any pair of vectors u, v.

Of supreme importance in geometry is the commutator of vector fields.

Example 10.4

The commutator is $[u, v]^j = \sum_i \left(u^i \partial_i v^j - v^i \partial_i u^j \right)$. In other words, $[A, B] = A \cdot \nabla B - B \cdot \nabla A$.

Exercise 10.1 Show that the commutator of vector fields satisfies the Jacobi identity.

Example 10.5

Vector fields of zero divergence preserve density and form another important example. That is, *the commutator of two vector fields with zero divergence also has zero divergence.* One way to see this is from the identity

$$[B,A] = \nabla \times (A \times B) - A\nabla \cdot B + B\nabla \cdot A.$$

We will return to this example, as it is central to fluid mechanics.

Example 10.6

The commutator of $n \times n$ matrices $AB - BA$ is another example of a Lie bracket. The Weyl basis consists of n^2 matrices e_j^i which have just one non-zero entry

$$\left[e_j^i\right]_k^l = \delta_k^i \delta_j^l, \quad e_j^i = \begin{bmatrix} 0 & 0 & \cdots & 0 & \cdots & 0 \\ 0 & 0 & \cdots & 0 & \cdots & 0 \\ 0 & 0 & \cdots & 1 & \cdots & 0 \\ 0 & 0 & \cdots & 0 & \cdots & 0 \\ 0 & 0 & \cdots & 0 & \cdots & 0 \\ 0 & 0 & \cdots & 0 & \cdots & 0 \end{bmatrix}.$$

Thus we get

$$e_j^i e_l^k = \delta_j^k e_l^i.$$

The commutation relations are

$$[e_j^i, e_l^k] = \delta_j^k e_l^i - \delta_l^i e_j^k.$$

Example 10.7

Anti-symmetric $n \times n$ matrices have commutators that are also anti-symmetric, forming the Lie algebra $so(n)$. It describes rotations in $\mathbb{R}^n$.

Anti-hermitean $n \times n$ matrices form a Lie algebra as well, called $u(n)$. These two are the best known examples of Lie algebras.

Exercise 10.2 Show that the set of traceless anti-hermitean matrices is a Lie algebra, called $su(n)$. Find a basis for $su(2)$ in terms of the Pauli matrices

$$\sigma_1 = \begin{pmatrix} 0 & 1 \\ 1 & 0 \end{pmatrix}, \quad \sigma_2 = \begin{pmatrix} 0 & -i \\ i & 0 \end{pmatrix}, \quad \sigma_3 = \begin{pmatrix} 1 & 0 \\ 0 & -1 \end{pmatrix},$$

familiar from quantum mechanics. How is $su(2)$ related to $so(3)$?

10.3.1 Structure constants

If is often convenient to write a vector in terms of its components w.r.t. a basis

$$v = v^k e_k,$$

for example, the Fourier components, or Cartesian components. The Lie bracket of basis vectors can also be so expanded

$$[e_k, e_q] = C^p_{kq} e_p.$$

(The sun on p is implied here.) Clearly

$$C^p_{kq} = -C^p_{qk}$$

from anti-symmetry. The Jacobi identity becomes

$$C^p_{kq} C^s_{pr} + C^p_{qk} C^s_{pr} + C^p_{kr} C^s_{pq} = 0.$$

For the cross product the structure constants vanish if any pair of indices coincide; otherwise,

$$C^3_{12} = 1 = C^1_{23} = C^2_{31}.$$

All the other components of C are given by using anti-symmetry.

Exercise 10.3 Choose $k = (k_1, k_2) \in \mathbb{Z}^2_\Lambda - \{0, 0\}$ (i.e., pairs of integers modulo Λ except for $(0,0)$):

$$[e_k, e_q] = \frac{\Lambda}{2\pi} \sin\left[2\pi \frac{k \times q}{\Lambda}\right] e_{k+q} \mod \Lambda.$$

Verify that this satisfies the Jacobi identity. What is the dimension? Show (McLachlan, 1993) (by a change of basis) that this is isomorphic to $su(\Lambda)$.

....

$su(\Lambda)$ arises naturally in quantum mechanics (Zachos et al., 2005). See below for an application to fluid mechanics.

10.3.2 Enveloping Poisson algebra of a Lie algebra

While we are at it we might as well review the concept of a Poisson algebra and two ways it is connected to a Lie algebra; see Section 7.2 of (Rajeev, 2013).

A *Poisson algebra* is a commutative algebra along with another bilinear operation (the *Poisson bracket*) satisfying the conditions

$$\{F, aG + bH\} = a\{F, G\} + b\{F, H\}, \quad a, b \in R, \quad \text{linearity}$$
$$\{F, G\} = -\{G, F\}, \quad \text{anti-symmetry}$$
$$\{\{F, G\}, H\} + \{\{G, H\}, F\} + \{\{H, F\}, G\} = 0, \quad \text{Jacobi identity}$$

and

$$\{F, GH\} = \{F, G\}H + F\{G, H\}. \quad \text{Leibnitz identity}$$

The first three conditions simply say that the Poisson bracket satisfies the conditions for a Lie bracket. This is the first connection between Poisson and Lie. The last condition is a compatibility of the commutative multiplication with the anti-symmetric bracket.

In most cases of interest, the elements of a Poisson algebra are functions on some manifold. Then the first and last conditions say that there is a tensor field ("Poisson tensor") r^{ij} such that

$$\{F, H\} = r^{ij} \partial_i F \partial_j H.$$

The remaining conditions then are properties of the Poisson tensor:

$$r^{ij} = -r^{ji}, \quad \text{anti-symmetry}$$
$$r^{il} \partial_l r^{jk} + r^{jl} \partial_l r^{ki} + r^{ki} \partial_l r^{ij} = 0. \quad \text{Jacobi identity}$$

We saw that every Poisson algebra is in particular a Lie algebra.

In the other direction, given a Lie algebra we can construct a Poisson algebra on its *envelope*, that is, the space of polynomial functions on the dual of the Lie algebra.

It is well known in Lie theory (Hausner and Schwartz, 1968; Serre, 1965) that this "universal envelope" of a Lie algebra is an associative algebra under a multiplication (sometimes called the star product) which is given as an expansion involving multiple Lie brackets. This is the Baker–Campbell–Hausdorff expansion or Poincaré–Birkhoff–Witt series. (It is such a wonderful thing that it was discovered and rediscovered by many people in different contexts.) The zeroth order term (with no Lie brackets in it) is the commutative product. The first order term is the Poisson bracket. This is all we need, mercifully.

You should always think of a Poisson algebra as an approximation to a non-commutative associative algebra. For a physicist, the way to understand this is in terms of quantum mechanics. The observables of quantum mechanics form such a non-commutative algebra. If we expand the product in powers of $\hbar$, to zeroth order we get the commutative product of functions on phase space. The first order correction is the Poisson bracket. Both are needed in the classical (i.e., small $\hbar$) limit. With this in mind, the universal enveloping algebra of Lie theory is the algebra of "quantum observables" while the enveloping Poisson algebra is the "classical limit." In the Baker–Campbell–Hausdorff expansion expansion, each appearance of a commutator counts as a power of $\hbar$. Not surprisingly, it is the Poisson algebra that we need in (classical) fluid mechanics.

If we expand an element of a Lie algebra in a basis $v = v^k e_k$, a (polynomial) function on the *dual* is determined by the coefficients (symmetric tensors) in the expansion

$$F(\xi) = \sum_{r=0} F^{i_1 \cdots i_r} \frac{\xi_{i_1} \cdots \xi_{i_r}}{r!}.$$

The multiplication of these functions corresponds to the symmetrized tensor products of their coefficients. The Poisson tensor is built out of the structure constants $r^{ij} = C^k_{ij} \xi_k$. Thus the Poisson bracket on polynomials is an extension of the underlying Lie bracket:

$$\{F, H\} = C^k_{ij} \xi_k \frac{\partial F}{\partial \xi^i} \frac{\partial H}{\partial \xi^j}.$$

The original Lie algebra is just the particular case of linear functions on its dual.

If there is a metric on the Lie algebra, we can identify the Lie algebra with its dual. You have to be careful doing this, as the metric may not be invariant under the Lie bracket (see below). This is the case of most interest to us, as we see in the next section. Still, understanding the connection of Lie and Poisson algebras in their natural language is beneficial.

10.3.3 Metric Lie algebra

A metric Lie algebra is a Lie algebra with an inner product. We say that the metric is *invariant* if

$$g([u, v], w) + g(v, [u, w]) = 0.$$

In terms of components,

$$g_{kl} C^l_{ij} + g_{jl} C^l_{ik} = 0.$$

The scalar and vector product in $\mathbb{R}^3$ is an example:

$$(u \times v) \cdot w = -v \cdot (u \times w).$$

(Prove this.) We are mostly interested in inner products that are *not* invariant: the equations we are interested in are completely trivial for the invariant case.

A metric Lie algebra defines a hamiltonian system. The idea is to think of the Lie algebra as the set of observables of the hamiltonian system and the Lie bracket as the Poisson bracket. The square of the length of a vector (its inner product with itself) is the hamiltonian (up to a constant factor). Together they define a dynamical system.

$$H = \frac{1}{2} g^{lm} u_l u_m,$$

$$\{u_l, u_k\} = C^n_{lk} u_n, \tag{10.2}$$

$$\frac{du_k}{dt} \equiv \{H, u_k\} = g^{lm} C^n_{lk} u_m u_n.$$

Exercise 10.4 Show that this r.h.s vanishes if g is an invariant metric.

Example 10.8

The rigid body is the basic example. The Lie algebra is that of rotations (cross product) and the covariant metric is the moment of inertia matrix. If the metric is proportional to the dot product, the equations of motion would be trivial. In this case the metric is invariant: the rigid body is isotropic, with equal moments of inertia in all direction. So, the interesting case is when the metric is not invariant.

The most interesting (but also most complicated) example for us is the incompressible fluid.

Example 10.9

Ideal fluid The Lie algebra is the commutator of vector fields with zero divergence. The metric is given by the L^2-inner product. The Euler equations in vorticity form follow from these. We will work our way up to this by working on lower dimensional (one and and two dimensional fluid flow) cases first.

Thus the Euler equations of both the rigid body and the ideal fluid describe the extent to which the kinetic energy is *not* invariant under the corresponding Lie algebra.

10.4 The Virasoro algebra

The simplest example of a vector field is that on a circle. Such a vector field has only one component, which must be a periodic function. In terms of this component, the commutator is just

$$[u, v] = uv' - vu'.$$

It is convenient to introduce a basis

$$e_m = -ie^{imx}\frac{\partial}{\partial x}, \quad m \in \mathbb{Z}.$$

By Fourier analysis, any vector field can be written as a sum

$$u = \sum_m u_m e_m.$$

Since u is real the coefficients must satisfy the reality condition

$$u_m^* = -u_{-m}.$$

The commutation relations are, in this basis,

$$[e_m, e_n] = (n - m)e_{m+n}.$$

Equivalently, this can be thought of as Poisson brackets among the components

$$\{u_l, u_m\} = (m - l)u_{l+m}.$$

This is sometimes called the Witt algebra.

Closely related to this algebra is the Virasoro algebra which has one extra generator c which commutes with all others:

$$\{u_m, u_n\} = (n - m)u_{m+n} + cm^3\delta(m + n),$$
$$\{u_m, c\} = 0.$$

The Witt and Virasoro algebras are examples of graded Lie algebras. Such algebras are central to modern theories of physics such as string theory.

....

Exercise 10.5 Verify the Jacobi identity for the Virasoro commutation relations above.

....

10.4.1   Burgers and KdV equations

We will now show that the celebrated KdV equation is a hamiltonian system based on the Virasoro algebra (Khesin and Wendt, 2009). We can choose as hamiltonian the total kinetic energy of a fluid of constant density moving in the circle:

$$H = \frac{1}{2} \int u^2(x) \frac{dx}{2\pi}.$$

(The 2π is put in to simplify the calculations: it amounts to choosing units such that the volume of the circle is one.) Geometrically, this defines a metric on the space of vector fields (L^2-norm). Using the structure constants of the Witt algebra of vector fields, we get some equations for the Fourier components. In the notation of eqn (10.2), $g^{lm} = \delta(l + m)$, $C_{lk}^n = (k - l)\delta(n - [k + l])$:

$$\frac{du_k}{dt} = \sum_{lmn} (k - l)\,\delta(l + m)\delta(n - [k + l])u_m u_n$$

$$= \sum_{mn} (k + m)\,\delta(n + m - k)u_m u_n$$

$$= \sum_{mn} (2m + n)\,\delta(n + m - k)u_m u_n.$$

Transforming to position space

$$u(x) = -i \sum_k u_k e^{ikx},$$

this becomes the inviscid Burgers equation

$$\frac{\partial u}{\partial t} = -3u \frac{\partial u}{\partial x}.$$

If we had used the Viraosoro algebra instead, there would be an extra generator that commutes with everything. We can choose as hamiltonian

$$H = \frac{1}{2} \int u^2(x) \frac{dx}{2\pi} + \frac{1}{2}c^2 = \frac{1}{2} \sum_m u_{-m} u_m + \frac{1}{2}c^2.$$

Since c has zero commutators, its derivative is zero: it is just a constant parameter. But it affects the equation for u by adding an extra term:

$$\frac{du_k}{dt} = \sum_{mn} (2m + n)\,\delta(n + m - k)u_m u_n + c(k^3 - k)u_k.$$

Transforming to position space we get an extra term involving the third derivative

$$\frac{\partial u}{\partial t} = -3u\frac{\partial u}{\partial x} - c\left[u''' + u'\right].$$

This equivalent to the KdV equation. We just have to identify

$$\phi(x) = au(x) + b$$

for appropriate constants a, b. The moral is that the Virasoro algebra along with the obvious definition of hamiltonian (kinetic energy) gives a natural hamiltonian formulation for the KdV equation. We already saw a different hamiltonian and commutation relation which also gives the same equation.

It is rare that a dynamical system has two such hamiltonian structures: KdV is an example of a "bi-hamiltonian" system. Such systems have an infinite number of conservation laws which can be deduced by hopping back and forth between the two hamiltonian formulations. This is one way to understand why the KdV equation is exactly solvable. You can spend a lifetime in the perfect world of integrable systems. But chaotic systems are closer to the real world of interest to physicists. We have to retrain our aesthetic sense to appreciate the beauty underlying chaos.

10.5   Hamiltonian for the two-dimensional Euler equations

Recall that an incompressible vector field can be written in terms of a velocity potential

$$\partial_1 v^1 + \partial_2 v^2 = 0 \implies v^1 = \partial_2\Omega, \quad v^2 = -\partial_1\Omega.$$

Thus, v is orthogonal to the gradient of Ω.

In two dimensions, vorticity is a scalar: the curl of any vector field in the plane points in the direction normal to the plane, and so can be treated as a scalar:

$$\omega = \partial_1 v_2 - \partial_2 v_1.$$

In fact

$$\omega = -[\partial_1^2 + \partial_2^2]\Omega.$$

If an ideal fluid fills some domain D, at the boundary the normal component of v must vanish: the fluid should not cross the boundary. Since $\nabla\Omega$ is orthogonal to v, the *tangential* component of $\nabla\Omega$ must vanish at the boundary, that is, Ω is constant along the boundary. Since adding a constant to Ω does not change v, we can choose the boundary value of Ω to be zero: the Dirichlet boundary conditions.

If we choose units in which density is equal to one, the energy of the fluid is

$$H = \frac{1}{2} \int_D v^2 d^2x = \frac{1}{2} \int \left[(\partial_1 \Omega)^2 + (\partial_2 \Omega)^2 \right] d^2x.$$

The two-dimensional Euler equations

$$\frac{\partial v}{\partial t} + v \cdot v = -\nabla p$$

can be written in vorticity form as

$$\frac{\partial \omega}{\partial t} + v \cdot \nabla \omega = 0.$$

Since ω points in the direction normal to the plane, $\omega \cdot \nabla v = 0$ and $[v, \omega] = v \cdot \nabla \omega$. Thus

$$\frac{\partial \omega}{\partial t} = \nabla \Omega \times \nabla \omega.$$

10.5.1   Poisson brackets

We can understand ideal incompressible flow as a Hamiltonian system, analogous to the rigid body. The vorticity is the basic observable, analogous to the angular momentum of a rigid body. The analog of angular velocity is the velocity potential Ω. The relation

$$\omega = -\nabla^2 \Omega$$

shows that $-\nabla^2$ is the analog of moment of inertia. Indeed recall that $-\nabla^2$ has positive eigenvalues and is a symmetric operator. The kinetic energy can written as

$$H = \frac{1}{2}(\Omega, \Delta\Omega) = \frac{1}{2}(\omega, \Delta^{-1}\omega),$$

where

$$(f, g) = \int f(x)g(x)d^2x,$$

again analogous to the rigid body. The inverse of the Laplacian is the Green's function with Dirichlet boundary conditions:

$$\Omega(x) = \Delta^{-1}\omega(x) = \int G(x, y)\omega(y)d^2y,$$

$$\Delta G(x, y) = \delta(x, y).$$

The commutator of two incompressible vector fields is again incompressible. Thus, if

$$u_1 = \partial_2 f, \quad u_2 = -\partial_1 f,$$
$$v_1 = \partial_2 g, \quad v_2 = -\partial_1 g,$$

there must be a function h such that

$$[u, v]_1 = \partial_2 h, \quad [u, v]_2 = -\partial_1 h.$$

In fact

$$h = \{f, g\} \equiv \nabla f \times \nabla g.$$

This operation $\{f, g\}$ satisfies the Jacobi identity. In fact, it is a version of the Poisson bracket familiar from mechanics: here x^1, x^2 are conjugate variables:

$$\{f, g\} = \partial_1 f \partial_2 g - \partial_2 f \partial_1 g.$$

10.5.2   Fourier basis

Perhaps the simplest case are periodic boundary conditions on the velocity potential:

$$\Omega(x_1 + L, x_2) = \Omega(x_1, x_2 + L) = \Omega(x_1, x_2).$$

For then we can expand in a Fourier series

$$\Omega(x) = \sum_{k \in \mathbb{Z}^2} \Omega_k e^{2\pi i \frac{k \cdot x}{L}}.$$

Here $k = (k_1, k_2)$ is a pair of integers. Then

$$\omega(x) = \left(\frac{2\pi}{L}\right)^2 \sum_{k \in \mathbb{Z}^2} k^2 \Omega_k e^{2\pi i \frac{k \cdot x}{L}}$$

and the Euler equations become

$$\frac{\partial \omega_k}{\partial t} = \sum_{p \in \mathbb{Z}^2}' \omega_p \omega_{k-p} \frac{p \times k}{p^2}.$$

(The prime on the summation symbol $\sum'$ means that the term $p = 0$ should be excluded.)

Because of the scale invariance of the Euler equations, L has dropped out. This is an infinite dimensional system of ODEs: p and k are pairs of integers.

10.6   Spectral discretization of two-dimensional Euler

It is often useful to approximate a set of differential equations with an infinite number of unknowns by one with a finite number of unknowns. This is necessary for numerical solution; also useful sometimes in studying theoretical issuess such as existence, regularity or stability.

A cutoff in our case would be to replace $\mathbb{Z}$ by $\mathbb{Z}_\Lambda$, that is, the set of numbers $\mathbb{Z}_\Lambda = \{0, 1, \cdots \Lambda\}$ where addition is modulo Λ. (For example, $10 + 8 = 6$ modulo 12.) Notice that now we have periodicity both in space and in the conjugate ("wavenumber") variable. To retain the periodicity, we must replace $p \times q$ and $p \cdot p$ by periodic functions (McLachlan, 1993).

$$p \times k \rightarrow A(p, k) \equiv \frac{\Lambda}{2\pi} \sin\left[2\pi \frac{p \times k}{\Lambda}\right],$$

$$p \cdot p \rightarrow \Delta(p) \equiv \frac{\Lambda^2}{\pi^2} \left\{\sin^2\left[\frac{\pi p_1}{\Lambda}\right] + \sin^2\left[\frac{\pi p_2}{\Lambda}\right]\right\}.$$

We chose these functions such that as $\Lambda \rightarrow \infty$ they reduce to the original quantities. Then we have

$$\frac{\partial \omega_k}{\partial t} = \sum_{p \in \mathbb{Z}_\Lambda^2}' \omega_p \omega_{k-p} \frac{A(p, k)}{\Delta(p)}. \tag{10.3}$$

This is a finite system of ODEs that can be solved numerically for values of $\Lambda \sim 10^4$ on a laptop computer. We can observe that in two dimensions, vortices tend to merge to form larger ones as time evolves.

....

Exercise 10.6 Write a program that solves eqn (10.3) numerically and plots the velocity field.

....

....

Exercise 10.7 Research project Generalize the above theory to fluid flow on a sphere. Instead of Fourier series, you must use spherical harmonics. With some ingenuity, you can find a deformaton of the Poisson algebra of functions on a sphere that is isomorphic to $su(n)$: a "star product" (Zachos et al., 2005).

....

If we postulate (McLachlan, 1993) the P.B. to form the Lie algebra $su(\Lambda)$ (we saw this Lie algebra earlier, in Exercise 10.3):

$$\{\omega_p, \omega_q\} = \frac{\Lambda}{2\pi} \sin\left[2\pi \frac{p \times k}{\Lambda}\right] \omega_{p+q}$$

and a hamiltonian

$$H = \frac{1}{2} \sum_{p \in \mathbb{Z}_\Lambda^2} \frac{|\omega_p|^2}{\Delta(p)},$$

we get the above ODE as the equations of motion. Coincidentally, when $\Lambda = 2$ the Lie algebra $su(2)$ is the same as that of rotations. Thus in this case the equations above are the Euler equations of a rigid body. Thus, as Λ varies from 2 to ∞, we interpolate between Euler equations for a rigid body to Euler equations for a two-dimensional ideal fluid!

Next we turn to the more realistic three-dimensional ideal fluid.

10.7   Clebsch variables

It is a theorem of vector analysis (Lamb, 1945) that any vector field in R^3 can be written as

$$u = \nabla\lambda + \nabla\alpha \times \nabla\beta.$$

The three components of u determine the scalar functions α, β, λ (up to some constants). For example, λ is determined by the divergence of u:

$$\nabla \cdot u = \nabla^2\lambda.$$

Thus, if a vector field has zero divergence, we can set $\lambda = 0$ in this decomposition. In fluid mechanics, this can be applied to vorticity (being the curl of velocity, it has zero divergence):

$$\omega = \nabla\alpha \times \nabla\beta.$$

These α, β are called Clebsch variables. The velocity of an incompressible fluid can also be determined in terms of α, β

$$v = \nabla\gamma + \frac{1}{2}[\alpha\nabla\beta - \beta\nabla\alpha].$$

γ is not a new degree of freedom: it is determined in terms of α, β by solving the Poisson equation

$$\nabla \cdot v = 0 \iff \nabla^2 \gamma = -\frac{1}{2} \nabla \cdot [\alpha \nabla \beta - \beta \nabla \alpha].$$

The Euler equations in vorticity form

$$\frac{\partial v}{\partial t} + [v, \omega] = 0$$

become in terms of these variables

$$\frac{\partial \alpha}{\partial t} + v \cdot \nabla \alpha = 0,$$

$$\frac{\partial \beta}{\partial t} + v \cdot \nabla \beta = 0.$$

Here, α and β are thought of as the basic dynamical variables and v is determined by them as above. The geometric meaning of these equations is this: α, β constants along the integral curves of v in space-time.

....

Exercise 10.8 Show that, if α, β satisfy the above equations, vorticity satisfies the appropriate Euler equations.

....

10.8   Hamiltonian form of 3D Euler equations

The main reference for this section is the paper by (Marsden and Weinstein, 1983). For a particle physicist's perspective see (Jackiw, 2002).

In units where the density is one, the energy of an ideal fluid is simply

$$H = \frac{1}{2} \int v^2 \, dx.$$

Recall the vorticity and velocity potential are defined by

$$\omega = \nabla \times v, \quad v = \nabla \times \Omega.$$

Using the identity

$$\nabla \cdot [v \times \Omega] = \Omega \cdot \nabla \times v - v \cdot \nabla \times \Omega$$

and dropping the integral of a total divergence, we get

$$H = \frac{1}{2} \int \Omega \cdot \omega dx.$$

In terms of Clebsch variables

$$H = \frac{1}{2} \int \Omega \cdot \nabla \alpha \times \nabla \beta dx.$$

Equivalent forms obtained by integration by parts are

$$H = \frac{1}{2} \int \alpha \nabla \cdot (\Omega \times \nabla \beta) = \frac{1}{2} \int \alpha v \cdot \nabla \beta dx.$$

Similarly,

$$H = -\frac{1}{2} \int \beta v \cdot \nabla \alpha dx.$$

Thus, the hamiltonian depends quadratically on each of the variables α, β. Thus

$$\frac{\delta H}{\delta \alpha} = v \cdot \nabla \beta,$$

$$\frac{\delta H}{\delta \beta} = -v \cdot \nabla \beta.$$

(The factor of $\frac{1}{2}$ in the hamiltonian gets canceled by the factor of two coming from its quadratic dependence on α and on β.) So the equations of motion for the Clebsch variables can be written as

$$\frac{\partial \beta}{\partial t} = -\frac{\delta H}{\delta \alpha},$$

$$\frac{\partial \alpha}{\partial t} = \frac{\delta H}{\delta \beta}.$$

These are Hamilton's equations where α is like "position" and β is like the conjugate canonical "momentum" variable. Thus α and β satisfy the canonical Poisson brackets

$$\{\beta(x), \alpha(y)\} = \delta(x, y), \quad \{\alpha(x, \alpha(y)\} = 0 = \{\beta(x), \beta(y)\}.$$

Thus, in a sense, the Clebsch variables are the canonical variables of an incompressible fluid. The complications of fluid mechanics arise because the hamiltonian is *quartic* in these variables. By contrast, simple systems (harmonic oscillator) have quadratic hamiltonians in canonical variables.

10.9   Poisson brackets of velocity

Let F be a function of the velocity field (such as the hamiltonian above). By an argument similar to the one above, we can show that

$$\frac{\delta F}{\delta \alpha} = \frac{\delta F}{\delta v} \cdot \nabla \beta,$$

$$\frac{\delta F}{\delta \beta} = -\frac{\delta F}{\delta v} \cdot \nabla \alpha.$$

Given a pair of such functions, their Poisson brackets are given by

$$\{F, G\} = \int \left(\frac{\delta F}{\delta \beta} \frac{\delta G}{\delta \alpha} - \frac{\delta F}{\delta \alpha} \frac{\delta G}{\delta \beta} \right) dx.$$

This is simply the usual formula for P.B. except that the sum over the indices on the canonical variables is replaced by an integral. Thus

$$\{F, G\} = \int \left(\frac{\delta F}{\delta v} \cdot \nabla \beta \frac{\delta G}{\delta v} \cdot \nabla \alpha - \frac{\delta F}{\delta v} \cdot \nabla \alpha \frac{\delta G}{\delta v} \cdot \nabla \beta \right) dx.$$

Using

$$(a \times b) \cdot (c \times d) = a \cdot c \, b \cdot d - a \cdot d \, b \cdot c$$

and $\omega = \nabla \alpha \times \nabla \beta$ we get

$$\{F, G\} = -\int \left[\frac{\delta F}{\delta v} \times \frac{\delta G}{\delta v} \right] \cdot \omega dx.$$

An integration by parts gives

$$\{F, G\} = \int v \cdot \nabla \times \left(\frac{\delta F}{\delta v} \times \frac{\delta G}{\delta v} \right) dx.$$

Using yet another vector identity this becomes

$$\{F, G\} = \int v \cdot \left[\frac{\delta F}{\delta v}, \frac{\delta G}{\delta v} \right] dx.$$

Thus the P.B. of fluid mechanics amount to the commutator of incompressible vector fields. (Recall that they form a Lie sub-algebra).

10.10   Analogy with angular momentum

This is similar to the formula for Poisson brackets of functions F, G of angular momentum:

$$\{F, G\} = L \cdot \frac{\partial F}{\partial L} \times \frac{\partial G}{\partial L},$$

which follows from the P.B. of angular momentum

$$\{L_i, L_j\} = \epsilon_{ijk} L_k.$$

A canonical realization of these P.B. is

$$L = r \times p,$$

where the position and momentum are canonically conjugate:

$$\{p_i, r_j\} = \delta_{ij}, \quad \{p_i, p_j\} = 0 = \{r_i, r_j\}.$$

The Clebsch parameters are canonical variables (analogues of r, p) for an ideal incompressible fluid. Nineteenth century physicists might have been tempted to think therefore that α, β have something to do with the microscopic degrees of freedom of a fluid. We now know that this is not the case: microscopically, a fluid is no more than a large number of particles. This is a cautionary tale for those looking for deep meaning underlying the nonlinear field theories of our time, such as general relativity or Yang–Mills theory.

11

Curvature and Instability

The Euler equations of a rigid body can be understood as determining geodesics on the rotation group, w.r.t. a Riemannian metric that is invariant under the left action of the group on itself (Rajeev, 2013). In a similar way, the Euler equations of an ideal fluid can be interpreted as the geodesic equation on the group of incompressible diffeomorphisms of R^3 (Arnold, 1966).

Arnold also showed that that the curvature is negative (in all but a finite number of directions) for the case of ideal fluids. For further developments see (Arnold and Khesin, 1998) and (Khesin et al., 2013). This gives a geometric explanation for the observed instability of fluid flow. Moreover, the magnitude of the curvature grows with higher wavenumber: a fluid is more unstable to high wavenumber perturbations. This is a possible mechanism for turbulence: a frontier of research in mathematical physics. However, a theory of turbulence has to include dissipation which is ignored in this discussion. In Section 11.5 we will see how to modify the geodesic equation to include dissipation.

Understanding the geometry of the diffeomorphism group is important for other deep problems of theoretical physics (e.g., quantum gravity). So we might as well use the opportunity afforded by fluid mechanics to take a deep dive into this topic. Be warned that this chapter is at a higher mathematical level than the rest of the book. Also, this chapter is not necessary to understand other chapters.

11.1   Riemannian geometry

Most physicists learn Riemannian geometry in the context of general relativity. It also has intimate connections to mechanics (Rajeev, 2013) and is a thriving branch of pure mathematics (Chavel, 2006; Lee, 1997). We recall the bare essentials here to set the context.[1]

---

[1] This chapter is based on some lectures at the Chennai Mathematics Institute.

Fluid Mechanics: A Geometrical Point of View. S. G. Rajeev © S. G. Rajeev 2018.
Published in 2018 by Oxford University Press. DOI: 10.1093/oso/9780198805021.001.0001

11.2   Covariant derivative

A covariant derivative (connection) $\nabla_u v$ of a vector field v on a manifold M along another vector field u should satisfy the conditions of linearity in u, v as well as

$$\nabla_u[fv] = f\nabla_u v + u(f)v, \quad \nabla_{fu}v = f\nabla_u v.$$

Thus, it involves first derivatives of v and no derivative of u. Explicitly in coordinates:

$$[\nabla_u v]^i = u^j \partial_j v^i + \Gamma^i_{jk} u^j v^k$$

for a set of *connection coefficients* Γ^i_{jk}. We will only be interested in connections without torsion:

$$\nabla_u v - \nabla_v u = [u, v] \iff \Gamma^i_{jk} = \Gamma^i_{kj}.$$

The curvature of such a connection is the tensor field defined by

$$\hat{r}(u, v)w = \nabla_u \nabla_v w - \nabla_v \nabla_u w - \nabla_{[u,v]} w. \tag{11.1}$$

⋯⋯⋯⋯⋯⋯⋯⋯⋯⋯⋯⋯⋯⋯⋯⋯⋯⋯⋯⋯⋯⋯⋯⋯⋯⋯⋯⋯⋯⋯⋯⋯⋯⋯⋯⋯

Exercise 11.1 Verify that that this quantity is a tensor field. That is, linearity in each argument over the ring of functions: $\hat{r}(fu, v)w = f\hat{r}(u, v)w$, $\hat{r}(u, v)fw = f\hat{r}(u, v)w$, etc.

Solution

$$\nabla_{fu}\nabla_v w - \nabla_v \nabla_{fu} w - \nabla_{[fu,v]} w = f\nabla_u \nabla_v w - \nabla_v(f\nabla_u w) - \nabla_{f[u,v]-v(f)u} w$$
$$= f\nabla_u \nabla_v w - f\nabla_v(\nabla_u w) - v(f)\nabla_u w - f\nabla_{[u,v]} w + v(f)\nabla_u w$$

etc.

⋯⋯⋯⋯⋯⋯⋯⋯⋯⋯⋯⋯⋯⋯⋯⋯⋯⋯⋯⋯⋯⋯⋯⋯⋯⋯⋯⋯⋯⋯⋯⋯⋯⋯⋯

In terms of components

$$r^i_{jkl} = \partial_j \Gamma^i_{kl} - \partial_k \Gamma^i_{jl} + \Gamma^i_{jm}\Gamma^m_{kl} - \Gamma^i_{km}\Gamma^m_{jl}.$$

Given a Riemannian metric g on the manifold, there is a unique connection of zero torsion and which preserves the metric:

$$g(\nabla_u v, w) + g(v, \nabla_u w) = u(g(v, w)).$$

Explicitly in coordinates, the connection coefficients are the *Christoffel symbols*

$$\Gamma^i_{jk} = \frac{1}{2}g^{il}\left\{\partial_j g_{lk} + \partial_k g_{lj} - \partial_l g_{jk}\right\}.$$

The curves that extremize the action

$$S = \frac{1}{2}\int g_{ij}(x)\dot{x}^i \dot{x}^j \, dt \tag{11.2}$$

are the geodesics. They satisfy the equation

$$\ddot{x}^i + \Gamma^i_{jk}\dot{x}^j \dot{x}^k = 0. \tag{11.3}$$

Equivalently,

$$\nabla_{\dot{x}}\dot{x} = 0,$$

where

$$\nabla_{\dot{x}}y^i \equiv \frac{dy^i}{dt} + \Gamma^i_{jk}\dot{x}^j \dot{x}^k$$

is the covariant derivative along the curve. The meaning of the geodesic equation is that the covraiant derivative of the tangent vector is zero along the curve. In particular, its length does not change.

....

Exercise 11.2 Derive the geodesic equation (11.3) as the condition for the above action (eqn 11.2) to be an extremum (i.e., show that the geodesic equation is its Euler–Lagrange equation. Show that the square of the length of the tangent vector $g_{ij}\dot{x}^i \dot{x}^j$ is a conserved quantity.

....

11.3 Geodesic deviation and curvature

The meaning of curvature is that it determines the rate of deviation of nearby geodesics from each other. This insight goes all the way back to Riemann's habilitation lecture, which is the founding document of the field.

Consider a geodesic $x^i(t)$, expressed in a coordinate system centered at the initial point: $x^i(0) = 0$. Let the initial velocity be $v \in T_0 M$. An infinitesimally close geodesic to this one will satisfy the Jacobi equation (obtained by taking the infinitesimal variation of the geodesic equation)

$$\nabla_{\dot{x}}^2 y^i + y^l r^i_{ljk} \frac{dx^j}{dt} \frac{dx^k}{dt} = 0. \tag{11.4}$$

The vector field y along the original geodesic connects the points at equal time on two nearby geodesics. It is called the Jacobi vector field.

Exercise 11.3 By differentiating eqn (11.3) in the direction y^i, derive Jacobi's geodesic deviation equation. This calculation can be simplified by using Riemann normal coordinates; but that would deprive you of the sheer joy of collecting the many terms in derivatives of the metric into the curvature tensor.

An equation for the length squared of the Jacobi vector field follows (Chavel, 2006):

$$\frac{1}{2} \frac{d^2 |y(t)|^2}{dt^2} = |\nabla_{\dot{x}} y|^2 - r(y, \dot{x}), \tag{11.5}$$

where

$$r(u, v) = r^i_{ljk} v^j v^k u^l u_i$$

is curvature contracted with two vectors (a *biquadratic*). Using eqn (11.1) we can write this as

$$r(u, v) = g\left(v, \hat{r}(v, u)u\right) = g\left(\nabla_{[u,v]}u, v\right) - g\left(\nabla_u \nabla_v u, v\right) + g\left(\nabla_v \nabla_u u, v\right). \tag{11.6}$$

Exercise 11.4 By contracting eqn (11.4) with y_i, derive eqn (11.5). Use the symmetry properties of the Riemann tensor to show that $r(v, v) = 0$.

Suppose the initial conditions for the Jacobi equation are

$$y(0) = 0, \quad \nabla_{\dot{x}} y(0) = u - v, \quad \dot{x} = v.$$

That is, we consider two geodesics starting at the same point but with slightly different initial velocities v and u. To first order the deviation is

$$y^i(t) = t \left[u^i - \dot{x}\right] + O(t^2).$$

The ansatz

$$|y(t)|^2 = t^2|u - v|^2 - \frac{t^3}{3}r(u, v) + \mathrm{O}(t^4)$$

solves the equation (11.5). The first term would have been the answer in Euclidean space (when curvature is zero). Thus, if $r(u, v) < 0$, an infinitesimal perturbation in the direction u to the geodesic with tangent vector v will grow with time: there is an instability. If $r(u, v) > 0$, the two geodesics would tend towards each other.

For example, on a sphere, two geodesics that start at the North Pole in different directions eventually come together at the South Pole. The sphere has positive curvature everywhere. On a hyperboloid on the other hand, geodesics always deviate away from each other. The hyperboloid has negative curvature everywhere.

11.4   Curvature as a bi-quadratic

The contracted version of curvature $r(u, v)$ actually contains all the information in the tensor r^l_{ijk}. It is useful to pause to understand such biquadratic forms. Let us begin with quadratic forms.

A covariant symmetric tensor is a bilinear map $Q : V \times V \to R$, satisfying $Q(u, v) = Q(v, u)$. On the other hand, a quadratic form is a function $Q : V \to R$ of a single vector that satisfies the scaling property

$$Q(\lambda u) = \lambda^2 Q(u).$$

Given a covariant symmetric tensor we can construct a quadratic by taking the special case when its entries are equal:

$$Q(u) = Q(u, u).$$

Conversely, we can get a symmetric tensor from a quadratic by "polarization:"

$$Q(u, v) = \frac{Q(u + v) - Q(u) - Q(v)}{2}.$$

Thus, a symmetric tensor is completely determined by its quadratic form. The classic example is the metric tensor itself.

In the same spirit, a tensor with the symmetries of the Riemannian curvature is fully determined by its biquadratic form. In the applications we have in mind, this is a much more convenient description, as we will be able to write explicit formulas more easily.

A biquadratic is a function on a vector space $T : V \times V \to \mathbb{R}$ that satisfies

$$T(au + bv, cu + dv) = (ad - bc)^2 T(u, v). \tag{11.7}$$

In particular,

$$T(au, v) = a^2 T(u, v), \quad T(u, v) = T(v, u), \quad T(u, u) = 0.$$

An example of a biquadratic is the square of the area of the parallelogram of sides u, v:

$$A(u, v) = g(u, u)g(v, v) - g(u, v)^2.$$

Given a fourth rank tensor tensor with the symmetries

$$T_{ijkl} = -T_{jikl} = -T_{ijlk} = T_{klij} \tag{11.8}$$

we get a bi-quadratic

$$T(u, v) = T_{ijkl}u^i u^l v^j v^k.$$

Conversely, a bi-quadratic determines a fourth rank tensor satisfying the above symmetries (eqn 11.8) through a polarization formula (Proposition II.1.1 of Chavel, 2006).

The curvature tensor has the symmetries of eqn (11.8) and hence is determined by the biquadratic (eqn 11.6).

The sectional curvature

$$k(u, v) = \frac{r(u, v)}{g(u, u)g(v, v) - g(u, v)^2}$$

only depends on the subspace defined by u, v. The subspace is unchanged by an invertible 2×2 matrix $\begin{pmatrix} a & b \\ s & d \end{pmatrix}$ acting on the pair u, v. The changes in the numerator and denominator of $k(u, v)$ cancel out. The sectional curvature $k(u, v)$ at a point is (Do Carmo, 1976) the gaussian curvature of the two-dimensional surface containing geodesics that start there and are tangential to the planed spanned by u, v.

11.5 Adding dissipation

The delightful geometry of geodesics describes an ideal world without dissipation. The conservation of kinetic energy $\frac{1}{2}g_{ij}\dot{x}^i \dot{x}^j$ might be spoiled in real world applications by the loss of energy due to friction. The simplest model of friction is a force proportional to velocity:

$$\ddot{x}^i + \Gamma^i_{jk}\dot{x}^j \dot{x}^k = -\gamma^i_j \dot{x}^j.$$

Instead of conservation of kinetic energy, we get

$$\frac{d}{dt}\left[\frac{1}{2}g_{ij}\dot{x}^i\dot{x}^j\right] = -\gamma_{ij}\dot{x}^i\dot{x}^j,$$

where

$$\gamma_{ij} = g_{ik}\gamma_j^k.$$

If γ_{ij} is anti-symmetric there is no loss of kinetic energy. The Lorentz force due to a magnetic field is of this type. To get dissipation, γ must be a positive symmetric tensor:

$$\gamma_{ij} = \gamma_{ji}.$$

Such a tensor is positive if its quadratic form

$$\gamma(u) = \gamma_{ij}u^i u^j$$

is always positive:

$$\gamma(u) \geq 0, \quad \gamma(u) = 0 \iff u = 0.$$

This means that all the eigenvalues of the symmetric matrix γ_{ij} are positive.

There is almost no work on the geometry of such "dissipative geodesics." The most obvious difference is that dissipative geodesics cannot be extended for ever.

....

Exercise 11.5 Take the simplest case of geodesics on $\mathbb{R}$ with the constant metric and dissipation:

$$\ddot{x} = -\gamma\dot{x}.$$

This could describe a particle moving in a viscous liquid like oil. A particle that starts at the origin with initial velocity v. What is the distance it has traveled as time $t \to \infty$?

....

....

Exercise 11.6 Research project Investigate how the stability of the dynamics of geodesics is affected by dissipation.

....

We will see that ideal fluid mechanics describes geodesics on an infinite dimensional Lie group. The kinetic energy, which is the quadratic form

$$g(v) = \frac{1}{2} \int v^2(x) dx,$$

determines a metric. Dissipation determines another quadratic form

$$\Gamma(u) = \frac{1}{2} \nu \int \left[\nabla_i v_j + \nabla_j v_i \right]^2 dx,$$

which determines another tensor field on the Lie group. Here, ν is kinematic viscosity.

11.6 Lie groups

Recall that a group $\mathfrak{G}$ is any set on which there is a multiplication that is

- associative $a(bc) = (ab)c$;
- has an identity $ae = a = ea$;
- every element has an inverse $aa^{-1} = e = a^{-1}a$.

Example 11.1

Recall that matrix multiplication is associative, with the identity matrix satisfying the second axiom above. But not every matrix is invertible. (The determinant has to be non-zero for the inverse to exist). The subset of invertible $n \times n$ matrices is a group, called the *general linear group*: $GL(n, \mathbb{R})$ if the entries are real and $GL(n, \mathbb{C})$ if they are complex.

A subgroup of a group is a subset which contains the identity as well as the product and inverses of its elements. Important subgroups of the general linear group are

- the orthogonal group $O(n) \subset GL(n, \mathbb{R})$ of matrices satisfying $AA^T = 1$. They describe rotations and reflections in $\mathbb{R}^n$. Here 1 denotes the $n \times n$ identity matrix.
- the complex analogue is the unitary group $U(n) \subset GL(n, \mathbb{C})$ of matrices satisfying $UU^\dagger = 1$

A *Lie group* is a group that is also a manifold (Hausner and Schwartz, 1968), that is, the group elements can be described by uniquely some (finite) set of number of coordinates. The multiplication and inverse must be differentiable functions of these coordinates. The matrix groups above are the most important Lie groups in physics, with applications in quantum mechanics, nuclear physics and particle physics. As manifolds they are finite dimensional.

....

Exercise 11.7 Show that the infinitesimal neighborhood of the identity in $O(n)$ consists of anti-symmetric matrices: $A = 1 + Ta$, $AA^T = 1 \implies a = -a^T$ to first order in t. What is the dimension of $O(n)$? Also, $A^{-1} = 1 - ta + t^2 a^2 + \cdots$ to second order in t.

Thus, relate the group commutator to the Lie bracket: $ABA^{-1}B^{-1} = 1 + [a, b] + \cdots$ up to second order.

....

The tangent space at the identity of a Lie group $\mathfrak{G}$ is a Lie algebra $\mathcal{G}$: an example of this is given in Exercise 11.7. Conversely, from every Lie algebra we can construct a Lie group to which it is tangent. This process of building a Lie group out of a Lie algebra uses a non-commutative version of the exponential function. It is at the heart of many solution methods for differential equations (Iserles, 2002).

11.7   Infinite dimensional Lie groups

For incompressible fluids, the configuration space is the group of volume preserving diffeomorphisms. It is infinite dimensional, so we can't directly use the tools of Riemannian geometry.

Many other advanced areas of theoretical physics also need infinite dimensional Lie groups: quantum field theory, general relativity, statistical physics. Their theory is in its infancy. Even the definition of an infinite dimensional manifold is not yet universally agreed upon.

One strategy is to approximate the beast by a finite dimensional manifold. The Lie algebra of area preserving vector fields can be approximated by $U(N)$. (We saw in Section 10.6 how this is done at the level of Lie algebras.) For three-dimensional fluid flow, we will pursue such a "regularization" as well. But in that case the finite dimensional approximation fails to be a Lie group, so some of the mathematical structure is lost.

In this chapter we pursue a different strategy. The Riemannian metric of interest on the group is invariant under the action of the group. In this case, we can reduce differential geometry to algebra on the tangent space at the identity. Thus, we can calculate quantities like curvature using algebraic methods. Infinite dimensional algebra is a lot simpler than geometry.

So we will need to learn how to reduce the differential geometry of a group to algebra. An excellent article by (Milnor, 1976) will help us.

11.8   Geometry of left-invariant metrics

A Lie algebra $\mathcal{G}$ can be thought of either as the tangent space at the identity or as the space of left invariant vector fields on a Lie group $\mathfrak{G}$. An inner product on $\mathcal{G}$ is thus equivalent to a left invariant Riemannian metric on $\mathfrak{G}$. We will study the geodesics and curvature of this metric. Using homogenity, all computations can be reduced to the Lie algebra. The basic reference is the article by (Milnor, 1976), especially Section 5. We will go a bit beyond Milnor in deriving explicit formulas in a form useful for applications to mechanics.

Given a curve $\gamma : [a, b] \to \mathfrak{G}$, its tangent vector at each point can be thought of as an element of the Lie algebra:

$$\frac{d\gamma}{dt} = \gamma v.$$

Thus the tangent vectors give a curve in the Lie algebra $v : [a, b] \to \mathcal{G}$, called a *hodograph* of the curve γ.

Given a positive inner product on the Lie algebra, we can define the action of the curve as

$$S(\gamma) = \frac{1}{2} \int_a^b G(v, v) dt = \frac{1}{2} \int G\left(\gamma^{-1}\dot{\gamma}, \gamma^{-1}\dot{\gamma}\right) dt.$$

This action is invariant under the left multiplication $\gamma(t) \mapsto h\gamma(t)$ by any constant $h \in \mathfrak{G}$. The extrema of this action are the geodesics on $\mathfrak{G}$ w.r.t. to a left invariant Riemannian metric determined by G.

The God of Notations is especially unkind to us here: too many things are called g. Knuth, the God of Fonts helps us out a bit, though. The group is $\mathfrak{G}$, the Lie algebra is $\mathcal{G}$, the Riemannian metric on the group is G. The Riemannian metric on a general manifold (which may not be a group) is called g.

Our aim is to translate the curvature, covariant derivative, etc. of the Riemannian geometry of left invariant metrics on the group to some quantities on the Lie algebra. The Lie algebra is much easier to generalize to infinite dimensions than the Lie group. The action principle is a good tool to do this translation.

11.9 Geodesics on a group manifold

A geodesic is a curve at which the action is stationary with respect to small variations. To compute this variation, let us consider a one-parameter family of curves; that is a map $\phi : [0, \epsilon] \times [a, b] \to \mathfrak{G}$, for some positive number ϵ. We require that the initial and final points are left unchanged, that is, $\phi(s, a)$ and $\phi(s, b)$ are independent of s.

Define

$$\frac{\partial \phi}{\partial t} = \phi v, \quad \frac{\partial \phi}{\partial s} = \phi u.$$

So the initial and final conditions on ϕ imply $u(s, a) = u(s, b) = 0$.

....

Exercise 11.8 Derive the *integrability condition*

$$\partial_s v = \partial_t u + [v, u].$$

Solution

$$\partial_t \left(\phi^{-1} \partial_s \phi \right) = -\phi^{-1} \partial_t \phi \phi^{-1} \partial_s \phi + \phi^{-1} \partial_t \partial_s \phi,$$

$$\partial_s \left(\phi^{-1} \partial_t \phi \right) = -\phi^{-1} \partial_s \phi \phi^{-1} \partial_t \phi + \phi^{-1} \partial_s \partial_t \phi.$$

Subtract these two equations. The last terms cancel. The remaining equation can be written in terms of u, v to get the identity required.

....

Regarding $\phi_s = \phi(s, .)$ for each value of s as a curve, the action becomes a function of s. Its derivative is

$$\partial_s S(\phi_s) = \int_a^b G(\partial_s v, v) dt = \int_a^b G\left(\partial_t u + [v, u], v \right) dt.$$

Define the linear operator $\tilde{v} : \mathcal{G} \to \mathcal{G}$ by

$$G\left(\tilde{v} w, u \right) = G\left([v, w], u \right) + G\left(w, [v, u] \right). \tag{11.9}$$

Thus, $\tilde{v}$ is the deformation of the metric by v: it would vanish if the metric were invariant. In particular,

$$G\left([v, u], v \right) = G\left(u, \tilde{v} v \right).$$

By integration by parts

$$\partial_s S(\gamma_s) = - \int_a^b G(u, \partial_t v - \tilde{v} v) dt.$$

Thus the geodesic equation on the group becomes the ordinary differential equation (ODE) on the Lie algebra:

$$\partial_t v - \tilde{v} v = 0.$$

This can be interpreted in terms of the covariant derivative of the tangent vector along itself:

$$D_v v = -\tilde{v} v, \quad \partial_t v + D_v v = 0. \tag{11.10}$$

We will need a more general formula for covariant derivative, $D_u v$ of vector along any other vector.

We will denote the covariant derivative on a group manifold by D to distinguish it from the covariant derivative on a general Riemannian manifold, which we denote by ∇. Similarly, an inner product on the Lie algebra (and the associated left-invariant metric on the Lie group) will be denoted by G, to distinguish from the generic metric g on a Riemannian manifold.

This will be helpful when we talk of the diffeomorphism group of a Riemannian manifold: the covariant derivative D on the group of diffeomorphisms and that on the underlying manifold ∇ are closely related, but not identical notions.

11.10　Covariant derivative on a group

The covariant derivative (Levi–Civita connection connection) is determined by the derivative of a left invariant vector field by another. The conditions of zero torsion and preserving the metric become, the case of a Lie group,

$$D_u v - D_v u = [u, v],$$
$$G\left(D_u v, w\right) + G\left(v, D_u w\right) = 0.$$

We will now solve these equations to get $D_u v$.

Taking the inner product of the first equation with w,

$$G\left(D_u v, w\right) - G\left(D_v u, w\right) = G\left([u, v], w\right).$$

Cyclic permutations give

$$G\left(D_v w, u\right) - G\left(D_w v, u\right) = G\left([v, w], u\right),$$
$$G\left(D_w u, v\right) - G\left(D_u w, v\right) = G\left([w, u], v\right).$$

Adding the first and third equations and subtracting the second, and using the invariance of the metric,

$$G\left(D_u v, w\right) = \frac{1}{2}\left\{G\left([u, v], w\right) - G\left([v, w], u\right) + G\left([w, u], v\right)\right\}. \tag{11.11}$$

Recalling the definition of eqn (11.9),

$$G\left(D_u v, w\right) = \frac{1}{2}G\left([u, v] - \tilde{u}v - \tilde{v}u, w\right).$$

Stripping away w,

$$D_u v = \frac{1}{2}\left\{[u, v] - \tilde{u}v - \tilde{v}u\right\}.$$

In particular

$$D_u u = -\tilde{u}u,$$

which agrees with the geodesic equation.

11.11   Curvature of a left-invariant metric

We will now calculate explicitly the curvature biquadratic for a metric Lie algebra, that is, for a Lie algebra $(\mathcal{G}, G)$ with an inner product on it:

$$G\left(D_v D_u u, v\right) = -G(D_u u, D_v v) = -G\left(\tilde{u}u, \tilde{v}v\right).$$

Similarly,

$$-G\left(D_u D_v u, v\right) = G\left(D_v u, D_u v\right) = \frac{1}{4}G\left([v, u] - \tilde{u}v - \tilde{v}u, \ [u, v] - \tilde{u}v - \tilde{v}u\right)$$

$$= -\frac{1}{4}|[u, v]|^2 + \frac{1}{4}|\tilde{u}v + \tilde{v}u|^2$$

using the abbreviation $G(u, u) = |u|^2$. Also, replacing $u \to [u, v], v \to u, w \to v$ in eqn (11.11) we get

$$G\left(D_{[u,v]} u, v\right) = \frac{1}{2}\{G\left([[u, v], u], v\right) - G\left([u, v], [u, v]\right) + G\left([v, [u, v]], u\right)\}$$

$$= \frac{1}{2}G\left([[u, v], u], v\right) - \frac{1}{2}|[u, v]|^2 + \frac{1}{2}G\left([v, [u, v]], u\right).$$

The curvature can be obtained by adapting eqn (11.1):

$$R(u, v) = G\left(D_{[u,v]} u, v\right) - G\left(D_u D_v u, v\right) + G\left(D_v D_u u, v\right)$$

$$= -\frac{3}{4}|[u, v]|^2 + \frac{1}{2}\{G\left([[u, v], u], v\right) + G\left([v, [u, v]], u\right)\} + \frac{1}{4}|\tilde{u}v + \tilde{v}u|^2 - G\left(\tilde{u}u, \tilde{v}v\right).$$

The middle term can be further simplified using

$$G([u, w], v) = G(\tilde{u}w, v) - G(w, [u, v])$$

and $G(\tilde{u}w, v) = G(\tilde{u}v, w)$. We get

$$\frac{1}{2}\{G\left([[u, v], u], v\right) + G\left([v, [u, v]], u\right)\} = |[u, v]|^2 + \frac{1}{2}G\left([u, v], \tilde{v}u - \tilde{u}v\right).$$

Thus

$$R(u, v) = \frac{1}{4}|[u, v]|^2 + \frac{1}{2}G([u, v], \tilde{v}u - \tilde{u}v) + \frac{1}{4}|\tilde{u}v + \tilde{v}u|^2 - G(\tilde{u}u, \tilde{v}v).$$

We can "complete the square" on the first two terms to get

$$R(u, v) = \frac{1}{4}|[u, v] + \tilde{v}u - \tilde{u}v|^2 + G(\tilde{u}v, \tilde{v}u) - G(\tilde{u}u, \tilde{v}v). \qquad (11.12)$$

Note that the sectional curvature in any plane thát contains a Killing vector is non-negative:

$$\tilde{u} = 0 \implies R(u, v) = \frac{1}{4}|[u, v] + \tilde{v}u|^2.$$

In particular, the curvature of a metric that is invariant under both left and right actions is positive. Such metrics (assumed to be of positive signature) exist only on compact Lie groups. The simplest example is $SU(2) = \mathbb{S}^{3}$ for which the bi-invariant metric is the standard metric on the 3-sphere.

11.12   Geodesics on *SO*(3)

Consider R^3 as a Lie algebra with the cross product as the Lie bracket:

$$[u, v] = (u_2 v_3 - u_3 v_2, u_3 v_1 - u_1 v_3, v_2 - u_2 v_1).$$

A corresponding Lie group is $SO(3)$. Any inner product in R^3 can be brought to the diagonal form by a rotation without changing the Lie bracket:

$$G(u, v) = G_1 u_1 v_1 + G_2 u_2 v_2 + G_3 u_3 v_3.$$

Thus

$$\begin{aligned} G(\tilde{u}v, w) = &G_1\{[u_2 v_3 - u_3 v_2]\, w_1 + [u_2 w_3 - u_3 w_2]\, v_1\} \\ &+ G_2\{[u_3 v_1 - u_1 v_3]\, w_2 + [u_3 w_1 - u_1 w_3]\, v_2\} \\ &+ G_3\{[u_1 v_2 - u_2 v_1]\, w_3 + [u_1 w_2 - u_2 w_1]\, v_3\} \end{aligned}$$

so that

$$[\tilde{u}v]_1 = \frac{G_1 - G_3}{G_1} u_2 v_3 + \frac{G_2 - G_1}{G_1} u_3 v_2, \cdots$$

The dots denote three more relations obtained by cyclic permutations. In particular,

$$[\tilde{v}v]_1 = \frac{G_2 - G_3}{G_1} v_2 v_3.$$

The geodesic equations become

$$\frac{dv_1}{dt} + \frac{G_3 - G_2}{G_1} v_2 v_3 = 0, \cdots$$

If we identify (G_1, G_2, G_3) with the principal moments of inertia and $(v_1, v_2, v_3) = (\Omega_1, \Omega_2, \Omega_3)$ with the components of angular velocity, the equations of motion for angular momentum $(L_1, L_2, L_3) = (G_1\Omega_1, G_2\Omega_2, G_3\Omega_3)$ become

$$\frac{dL_1}{dt} + \left[\frac{1}{G_2} - \frac{1}{G_3}\right] L_2 L_3 = 0$$

etc. Thus the Euler equations of a rigid body are geodesics on $SO(3)$ with a left-invariant metric determined by the moment of inertia.

Calculating as above gives the formula for curvature

$$R(u, v) = \frac{(G_2 - G_3)^2 + 2G_1(G_2 + G_3) - 3G_1^2}{4G_1} (u_2 v_3 - v_2 u_3)^2 + \cdots$$

In particular, if we choose $u = (0, \frac{1}{\sqrt{G_2}}, 0)$ to be a unit vector in the second principal direction and v to be a unit vector in the third direction the sectional curvature of the 23-plane is

$$K_{23} = \frac{(G_2 - G_3)^2 + 2G_1(G_2 + G_3) - 3G_1^2}{4G_1 G_2 G_3}.$$

The others are given by cyclic permutations.

11.13   The diffeomorphism group

It is a remarkable fact that the set of diffeomorphisms of a Riemannian manifold is itself a Riemannian manifold. This higher Riemannian geometry is the proper language of many interesting physical systems such as ideal fluids. Such repetitions of structures at a higher level happen quite often in mathematics: the set of Riemannian metrics on a manifold is itself a Riemannian manifold; the set of complex structures is a complex manifold; the set of Kähler structures is itself a Kähler manifold and so on.

$GL(3, \mathbb{R})$ is the set of invertible linear transformations in $\mathbb{R}^n$. Its subgroup $O(3)$ is the group of rotations and reflections. It describes the configurations of a rigid body: the set of transformations of points that preserves the origin as well as the distance between any pair of points.

A fluid allows for greater freedom: distances between points do not need to be preserved. The relevant group is the set of all differentiable maps

$$x \rightarrow \phi(x).$$

If we follow one such transformation by another the next effect will be the composition

$$\phi \circ \psi(x) = \phi(\psi(x)).$$

This is an associative operation:

$$\phi \circ (\psi \circ \chi) = (\phi \circ \psi) \circ \chi.$$

The identity map (which leaves every point unchanged) is the identity operation.

The inverse of ϕ would be a smooth map satisfying

$$\phi^{-1}(\phi(x)) = x$$

for all x. Again, not every smooth function $x \mapsto \phi(x)$ has an inverse. For example, the derivative matrix $\frac{\partial \phi^i}{\partial x^j}$ must have an inverse. (Unlike in the case of linear maps, this is not sufficient anymore.)

The set of smooth maps with smooth inverses forms a group, called the diffeomorphism group $\mathfrak{D}$. It is an infinite dimensional Lie group: to fully describe a smooth function we would need an infinite number of parameters.

Recall that the volume of an infinitesimal region near x is transformed by a smooth function:

$$dx \mapsto \det \frac{\partial \phi}{\partial x} dx,$$

where $\det \frac{\partial \phi}{\partial x}$ is the Jacobian determinant. The subgroup of diffeomorphisms *SDiff* that preserve the volume element,

$$SDiff = \left\{ \phi \in Diff \mid \det \frac{\partial \phi}{\partial x} = 1 \right\},$$

is of special interest in fluid mechanics: it is the configuration space of an incompressible fluid.

We saw earlier that the infinitesimal diffeomorphisms are vector fields; the commutator of vector fields turns it into a Lie algebra.

11.14   Lie algebra of vector fields

Recall that the space of vector of fields V on a manifold form a Lie algebra. The subspace of vector fields SV of zero divergence form a sub-algebra. These are the infinitesimal forms of the Lie groups *Diff* and *SDiff*, respectively.

The exponential map which turns a curve in the Lie algebra to one in the Lie group has an important meaning in fluid mechanics: it is the transformation (eqn 1.2) from the Euler picture to the Lagrange picture.

....

Exercise 11.9 Suppose $\phi(x) = x + \epsilon u(x)$ and $\psi = x + \epsilon v(x)$ are two diffeomorphisms close to the identity. Calculate $\phi^{-1}(x), \psi^{-1}(x)$ and $\phi \circ \psi \circ \phi^{-1} \circ \psi^{-1}$ to second order in ϵ. Show that

$$\psi \circ \phi \circ \psi^{-1} \circ \phi^{-1}(x) = x + \epsilon^2[u, v] + O(\epsilon^3).$$

Solution

$$\phi(\phi^{-1}(x)) = x \implies \phi^{-1}(x) + \epsilon u\left(\phi^{-1}(x)\right) = x \implies \phi^{-1}(x) = x - \epsilon u\left(\phi^{-1}(x)\right).$$

Iterating twice and keeping up to second order terms

$$\phi^{-1}(x) = x - \epsilon u\left(x - \epsilon u\left(x - \epsilon u(\phi^{-1}(x))\right)\right) = x - \epsilon u(x) + \epsilon^2 u \cdot \nabla u + O(\epsilon^3).$$

Now expand $\psi \left(\phi \left(\psi^{-1} \left(\phi^{-1}(x)\right)\right)\right)$ to second order in ϵ.

....

11.15   Diffeomorphisms of the circle

The simplest manifold is the circle. So the first example of a diffeomorphism group must be $Diff(S^1)$. The standard metric on the circle leads to the L^2-metric on this group:

$$G(u, v) = \int_0^{2\pi} u(x)v(x)dx.$$

The deformation tensor of a vector field on the circle is easily found from

$$G(\tilde{u}v, w) = G(uv' - vu', w) + G(v, uw' - wu') = \int \left\{ uv'w - vu'w + vuw' - vwu' \right\} dx$$

to be

$$\tilde{u} = -3u'.$$

Thus the covariant derivative on the Lie algebra is

$$D_u v = \frac{1}{2}\{[u, v] - \tilde{u}v - \tilde{v}u\} = u'v + 2uv'.$$

The geodesic equation is the inviscid Burger's equation

$$\frac{\partial v}{\partial t} + 3v\frac{\partial v}{\partial x} = 0.$$

(The factor of three can be removed by rescaling $v \to \frac{1}{3}v$).
  The curvature form

$$R(u, v) = \frac{1}{4}|[u, v] + \tilde{v}u - \tilde{u}v|^2 + G(\tilde{u}v, \tilde{v}u) - G(\tilde{u}u, \tilde{v}v)$$

reduces to

$$R(u, v) = |[u, v]|^2 = \int (uv' - vu')^2 \, dx.$$

Thus the L^2-metric on $Diff(S^1)$ has positive curvature. This is understandable physically: we saw that the geodesics (solutions of the Burgers equation) have a tendency to "focus" and create shocks.

11.16   The L^2-metric for vector fields on $\mathbb{R}^3$

The Lie algebra of vector fields on $\mathbb{R}^3$ admits the inner product

$$G(u, v) = \int u(x) \cdot v(x)dx$$

obtained by combining the Euclidean inner product on vectors at each point with an integration. The deformation tensor is

$$G(\tilde{u}v, w) = G(u \cdot \partial v - v \cdot \partial u, w) + G(v, u \cdot \partial w - w \cdot \partial u)$$

$$= \int \left\{ \left(u^i \partial_i v^j\right) w^j - \left(v^i \partial_i u^j\right) w^j + v^j \left(u^i \partial_i w^j\right) - v^j \left(w^i \partial_i u^j\right) \right\} dx$$

$$= \int \left\{ u^i \partial_i v^j - v^i \partial_i u^j - \partial_i \left(v^j u^i\right) - v^i \partial_j u^i \right\} w^j dx$$

$$= \int \left\{ -v^i \partial_i u^j - v^j \partial_i u^i - v^i \partial_j u^i \right\} w^j dx$$

$$= \int \left\{ -\partial_i u^j - \delta_{ij} \partial_k u^k - \partial_j u^i \right\} v^i w^j dx.$$

Thus

$$[\tilde{u}]_{ij} = -\delta_{ij}\partial_k u^k - \partial_i u_j - \partial_j u_i.$$

In particular it vanishes for a Killing vector (for which $\partial_i u_j + \partial_j u_i = 0$). The geodesic equation is

$$\partial_t u_i = \left[-\delta_{ij}\partial_k u^k - \partial_i u_j - \partial_j u_i\right] u^j$$

$$= -u_i\partial_k u^k - u^j\partial_j u - \frac{1}{2}\partial_i |u|^2.$$

In other words

$$\frac{\partial u}{\partial t} + u \cdot \nabla u + \text{div} u\ u + \frac{1}{2}\nabla |u|^2 = 0.$$

This is the "template matching equation" useful in algorithms for computer vision (Khesin et al., 2013; Mumford, 1998).

11.17   Incompressible diffeomorphisms

Given a scalar density on a manifold, the set of diffeomorphisms that preserve it form a subgroup:

$$\mathfrak{D}_\rho = \{\phi \in \mathfrak{D} | \rho(x) = \det \partial\phi(x)\rho\ (\phi(x))\}.$$

Its Lie algebra is the set of vector fields of zero divergence. Recall that the divergence of a vector field with respect to a density ρ is

$$\text{div} u = \frac{1}{\rho}\partial_i \left[\rho u^i\right].$$

From the identity

$$\text{div}[u, v] = u\,[\text{div}\ v] - v\,[\text{div}\ u]$$

it follows that incompressible (divergenceless) vector fields form a sub-Lie algebra, which we will call $\mathcal{G}$.

Given a Riemannian metric g on M there is an inner product (L^2) on the space of vector fields on M:

$$G(u, v) = \int g(u, v)\rho dx.$$

Define now the *deformation tensor* $\tilde{u}$ of a vector field:

$$G(\tilde{u}v, w) = \int \{g([u,v],w) + g(v,[u,w])\}\,\rho$$
$$= \int \{g(\nabla_u v - \nabla_v u, w) + g(v, \nabla_u w - \nabla_w u)\}\,\rho,$$

where ∇_u is the Riemannian covariant derivative on the underlying manifold (M,g). Thus

$$G(\tilde{u}v, w) = \int \{\nabla_u [g(v,w)]\}\,\rho - \int v^i w^j \left[\nabla_i u_j + \nabla_j u_i\right]\rho.$$

After an integration by parts the first term is zero, when u has zero divergence. It follows that

$$[\tilde{u}v]_j = -\left[\nabla_i u_j + \nabla_j u_i\right]v^j + \nabla_i \phi(u,v),$$

where $\phi(u,v)$ is to be chosen such that the l.h.s has zero divergence:

$$\nabla^i \left\{-\left[\nabla_i u_j + \nabla_j u_i\right]v^j + \nabla_i \phi(u,v)\right\} = 0.$$

Thus, absorbing $\nabla(g(u,v))$ into the gradient term,

$$[\tilde{u}v + \tilde{v}u] = -[\nabla_v u + \nabla_u v] + \nabla_i p(u,v),$$
$$[D_u v] = \frac{1}{2}\{[u,v] + \nabla_v u + \nabla_u v\} + \nabla p(u,v).$$

But

$$[u,v] = \nabla_u v - \nabla_v u$$

so that

$$[D_u v] = \nabla_v u + \nabla p(u,v), \quad \nabla^2 p(u,v) + \mathrm{div}\nabla_v u = 0.$$

In particular

$$D_v v = \nabla_v v + \nabla p,$$

with p chosen such that the divergence is zero. It follows that *the geodesic equation on the group of volume preserving diffeomorphisms is the Euler equation, as promised.*

The L^2-inner product allows us to split the space of vector fields into the subspace of divergence-free vector fields (which is also a sub-algebra) and another subspace of gradients:

$$u = u^T + \nabla\phi(u), \quad \text{div } u^T = 0, \quad \nabla^2\phi(u) = \text{div } u.$$

This is an orthogonal decomposition. The Riemannian covariant derivative in the group of volume preserving diffeomorphisms is just the transverse projection of the covariant derivative on (M, g):

$$D_u v = [\nabla_u v]^T.$$

Thus we can split the L^2-inner product on vector fields into transverse and longitudinal pieces:

$$G(u, v) = T(u, v) + S(u, v),$$

where

$$T(u, v) = G(u^T, v^T), \quad S(u, v) = G(\nabla\phi(u), \nabla\phi(v)).$$

In other words, even if w is not of zero divergence,

$$G(D_u v, w) = T(\nabla_u v, w).$$

11.18   Curvature of the diffeomorphism group

Thus, for the incompressible diffeomorphism group,

$$
\begin{aligned}
R(u, v) &= G\left(D_{[u,v]}u, v\right) - G\left(D_u D_v u, v\right) + G\left(D_v D_u u, v\right) \\
&= G\left(D_{[u,v]}u, v\right) + G\left(D_v u, D_u v\right) - G\left(D_u u, D_v v\right) \\
&= G\left(\nabla_{[u,v]}u, v\right) + T\left(\nabla_v u, \nabla_u v\right) - T\left(\nabla_u u, \nabla_v v\right).
\end{aligned}
$$

We use the fact that v has zero divergence.

Now,

$$\nabla_{[u,v]}u = r(u, v)u + \nabla_u \nabla_v u - \nabla_v \nabla_u u,$$

$$R(u, v) = \int r(u, v)\rho + G\left(\nabla_u \nabla_v u - \nabla_v \nabla_u u, v\right) + T\left(\nabla_v u, \nabla_u v\right) - T\left(\nabla_u u, \nabla_v v\right).$$

Set $w = \nabla_v u$, which may not have zero divergence, even though u and v are of zero divergence. Then,

$$G(\nabla_u w, v) = \int g(\nabla_u w, v)\, \rho$$

$$= \int \nabla_u [g(w, v)]\, \rho - \int g(w, \nabla_u v)\, \rho$$

$$= -G(w, \nabla_u v) = -S(w, \nabla_u v) - T(w, \nabla_u v),$$

and similarly for $G(\nabla_v \nabla_u u, v)$. Thus

$$R(u, v) = \bar{r}(u, v) + S(\nabla_u u, \nabla_v v) - S(\nabla_v u, \nabla_u v),$$

where

$$\bar{r}(u, v) = \int r(u, v)\rho dx$$

is the curvature form of (M, g) averaged by the density. Of course, on $\mathbb{R}^3$ $r(u, v) = \bar{r}(u, v) = 0$.

It is not difficult now to work out explicit answers for a flat torus and recover Arnold's original results. We find the present form simpler as well as more general.

This beautiful picture of ideal fluid flow as geodesics ought to lead to better algorithms for numerical solution. For ODEs there is a sophisticated and enormously useful geometric approach; see chapter 15 as well as (Iserles, 2002). It should be important to extend the techniques to partial differential equations (PDEs).

It is also important to see how this story is modified by dissipation, extending our discussion in Section 11.5 to infinite dimensions. Dissipative evolution of vector fields could also improve convergence of numerical methods used in computer vision (Mumford, 1998).

12

Singularities

The Cauchy problem for the incompressible Navier–Stokes (NS) equations is central to mathematical fluid mechanics. That is,

- Given initial data $v(x, 0)$ with div $v = 0$, does there exist a solution $v(x, t)$ to the incompressible NS equations (3.1)?
- If the initial data is smooth, is the solution smooth as well?
- Is the solution unique?

The foundations were laid by (Leray, 1934) in his Ph.D. thesis. Even after decades this paper is still worth reading (look out for some archaic terminology). While he was a prisoner of war, Leray pretended to be a topologist so that he would not be forced to work on a German war project using fluid mechanics. He laid the foundations of a completely different subject: spectral sequences in homotopy theory. His ruse worked: Leray is now more famous as a topologist.

Leray was not able to settle the questions above. In fact, they still remain open and are considered among the deepest problems in all of mathematics. But what he accomplished is monumental.

- If space were two-dimensional, there exists a unique smooth solution given smooth initial data (Ladyzhenskaya, 1987). The physically interesting case of three dimensions is the true challenge.
- There exists a weak solution (i.e., in the sense of a distribution; see below) for all time given regular initial data. This weak solution may not be unique, however.
- If the initial data is smooth, this solution is smooth as well at least up to some finite time $\tau = \frac{Av}{V^2(0)}$, where A is a constant (which may depend on the data) and $V(0) = \max_x |v(x, 0)|$ is the maximum speed of the initial data.
- For the time interval $[0, \tau]$ where it is regular, the solution is unique.
- If there are singularities (still unknown), they are confined to a set of measure zero in time. That is, the set of times at which a singularity arises are "rare." At these

Fluid Mechanics: A Geometrical Point of View. S. G. Rajeev © S. G. Rajeev 2018.
Published in 2018 by Oxford University Press. DOI: 10.1093/oso/9780198805021.001.0001

singular times, we could lose uniqueness: the solution might branch into several possibilities.

- As we approach the first singular time (if there is one) the speed must diverge at some point in space $|v(x, t)| > \frac{Av}{\sqrt{T-t}}$.

In the other direction, Leray tried to construct a singular solution. His guess was that if a singularity develops, it would be self-similar (scale invariant):

$$v(x, t) = \sqrt{\frac{a}{T-t}} U\left(\frac{x}{\sqrt{a(T-t)}}\right).$$

For this ansatz, NS equations reduce to a static system for $U(x)$ (time is eliminated). If there is a *regular* solution to that system, the NS equations would have a *singular* solution. Recent work (Necas et al., 1996; Tsai, 1998) rules out this possibility: that is, there are no self-similar singularities. There could still be singularities that have discrete scale invariance; or which are not scale invariant at all. Research on this continues (Seregin, 2015; Tao, 2016).

What would be the physical meaning of a singularity in a solution to NS equations? Recall that the NS equations are approximations, describing the average motion over a region of larger size than the mean free path of molecules. If a singularity develops, this approximation breaks down. A more accurate equation keeping the next order in the expansion in powers of inverse mean free path (Chapman and Cowling, 1970) will be needed. Such an equation would involve higher order spatial derivatives and will have smooth solutions for smooth data.

Even if there are singularities, it is unlikely that they are the source of the phenomenon of turbulence. We know that chaos can arise even in perfectly regular systems (even iterations of functions of a single variable). The question of singularities in NS equation is of conceptual and mathematical interest, but less crucial to physics.

Now we turn to a more in-depth description of the analysis of NS equations. Our aim is modest: to give a flavor of this highly technical subject, using physical concepts as much as possible. It is not possible to do much more than that in a book aimed at physicists.

12.1 Norms: L^2, L^p, Sobolev

In studying the solutions to any partial differential equation (PDE), it is important to have some control on the "size" of the solution. More precisely, on how far it deviates from the trivial solution (in our case, $v = 0$). Since the set of vector fields is linear, we must use some norm. The most obvious one is the L^2-norm

$$||v|| = \sqrt{\int v_i v_i(x)\, dx}.$$

This simply generalizes the notion of length of a vector in Euclidean space to the infinite dimensional space of vector fields, replacing the sum over components by an integral over space. Its physical significance for incompressible flow is obvious:

$$\frac{1}{2}\rho||v||^2 = \frac{1}{2}\int \rho v_i v_i(x) dx$$

is the total kinetic energy of the fluid. As such we have immediate control on its time evolution: we know that it cannot increase with time. The "size" of the velocity field, as measured by its L^2-norm will always decrease in time. If we could conclude from this that the velocity $v(x,t)$ itself remains bounded, there would have been no problem in ruling out singularities in NS evolution. Alas, the energy or L^2-norm does not always give sufficient control.

The quantity $||v||$ has dimensions of $\frac{L^{1+\frac{d}{2}}}{T}$ in d-dimensional space. So there is something special about $d=2$: it has the same units as kinematic viscosity in that case. Thus, the L^2-norm (or energy) is a scale invariant quantity under the scale symmetries that leave the NS equations invariant when $d=2$. Along with the dissipation inequality this gives sufficient control that solutions to two-dimensional NS equations with regular initial data remain regular.

For $d=3$, the case of most physical interest, we need finer measures of the size of a vector field. The L^p-norm is

$$||v||_p = \left[\int \{v_i v_i(x)\}^{\frac{p}{2}} dx\right]^{\frac{1}{p}}.$$

When $p=2$ it reduces the norm we discussed earlier. It scales like $\left[\frac{L^p}{T^p}L^d\right]^{\frac{1}{p}} = \frac{L^{1+\frac{d}{p}}}{T}$. Thus when $d=p$ it has dimensions of kinematic viscosity. The L^3-norm is scale invariant in dimension three. Unfortunately there is no simple inequality that says that the L^3-norm must decrease with time.

The L^p-norms are sensitive to size of the velocity field and the volume over which has support (i.e., it is non-zero). The larger the power p, the more the contribution of regions of large speed. In the limit $p \to \infty$, we get the L^∞-norm, which is simply the maximum speed over all of space (the supremum if you want to be analyst about it):

$$V = \sup_x \sqrt{v_i(x)v_i(x)}.$$

Leray's results show that if a singularity is to develop in a solution to NS equations, this quantity has to diverge.

Often we will need to control how rapidly the field varies in space. An example is the dissipation integral

$$\mathcal{J} = \left[\int (\partial_j v_i + \partial_i v_j)(\partial_j v_i + \partial_i v_j) d^3x\right]^{\frac{1}{2}}.$$

If we exclude translations and rotations (e.g., by requiring that $v_i(x)$ vanish at infinity) this is a norm as well.

This plays a crucial role in fluid mechanics because it measures the rate at which kinetic energy is lost. Generalizing this idea, we can consider a combination of higher derivatives

$$\mathcal{I}_m = \left[\int \frac{\partial^m u_i}{\partial x^{i_1} \cdots \partial x^{i_m}} \frac{\partial^m u_i}{\partial x^{i_1} \cdots \partial x^{i_m}} dx \right]^{\frac{1}{2}}.$$

This *Sobolev norm* plays a less crucial role in fluid mechanics.

The space of smooth vector fields is not complete with respect to any of these norms. That is, there are Cauchy sequences of smooth vector fields (whose distance from each other tend to zero) which do not tend to a smooth limit. This is similar to their being sequences of rational numbers that approach each other whose limit is not rational. Think of a sequence that approximates $\sqrt{2}$:

$$1.4, \quad 1.41, \quad 1.414 \quad 1.4142, \cdots, \quad 1.414213562, \cdots$$

To do any sort of sensible analysis, we have to fill in these "holes:" take the completion of the space with the norm. In the case of rational numbers this leads to the space of real numbers: each such number can be approximated as close as needed by rational numbers, although most are not rational themselves. In the same way, we can define vector spaces L^p, H^m obtained by completing the space of smooth vector fields by the L^p or Sobolev norm above. A standard approach to constructing solutions to a PDE is to first prove existence in one of these spaces: weak solutions. But there is a problem: most of the elements in these completed spaces are not smooth functions (just as most real numbers are not rational): instead they are generalized functions (also called distributions). To prove that the solutions are regular (that they are smooth fields) further work is needed.

12.2 The dissipation of energy

The analytical meaning of dissipation of energy, eqn (3.3), can now be made clear. For a smooth solution of NS, the L^2-norm must decrease at a rate given by the the Sobolev norm:

$$\frac{d}{dt}||v||^2(t) = -\nu\mathcal{I}^2(t).$$

By integrating both sides we get

$$||v||^2(t_0) = \nu \int_{t_0}^{t} \mathcal{I}^2(s)ds + ||v||^2(t).$$

This relation says that the initial energy is equal to the current energy plus the total dissipation up to the present time.

12.3   Solution of Navier–Stokes by perturbation theory

In the limit of slow motion (equivalently, large viscosity or small Reynolds number) we can ignore the nonlinear term $v \cdot \nabla v$, that is, reduce the Navier–Stokes (NS) to the Stokes equation. Being just the diffusion equation for each component of velocity, it has an explicit solution in terms of the diffusion kernel (also called the heat kernel). If the initial data is not too far from zero, we can attempt to treat the nonlinear term in a perturbation theory. The solution to NS is then constructed as an infinite series. Leray studies the convergence of this series systematically, and shows that it converges up to a finite time τ. The solution might continue for $t > \tau$: it just cannot be constructed by this method.

The Stokes equation

$$\frac{\partial v_i}{\partial t} = \nu \nabla^2 v_i - \frac{1}{\rho} \partial_i p, \quad v(x, 0) = u_i(x), \quad \partial_i u_i = 0$$

can be solved in terms of the diffusion kernel:

$$v_i(x, t) = \int K(x - y \mid t)\, u_i(y) dy, \quad p = 0,$$

$$K(x \mid t) = \frac{e^{-\frac{x^2}{4\nu t}}}{[4\pi \nu t]^{\frac{3}{2}}}.$$

You can check by integration by parts that divergence of v_i is zero. Since the pressure p is simply present to enforce the incompressibility, in the Stokes approximation it is constant (can be set to zero).

The inhomogenous Stokes equation has a forcing term on the r.h.s.:

$$\frac{\partial v_i}{\partial t} = \nu \nabla^2 v_i - \frac{1}{\rho} \partial_i p + f_i(x, t), \quad v(x, 0) = u(x).$$

The condition $\partial_i v_i = 0$ determines the pressure at each time:

$$\nabla^2 p = \rho \partial_i f_i.$$

It can also be solved in terms of the diffusion kernel:

$$v_i(x, t) = \int K(x - y \mid t)\, u(y) dy + \int_0^t ds \int K(x - y \mid t - s) \left[-\frac{1}{\rho} \partial_i p(y, s) + f_i(y, s) \right] dy.$$

....

Exercise 12.1 Use Fourier analysis to solve the inhomogenous Stokes equation.

....

By taking the nonlinear term in the NS equation to the r.h.s. and treating is as the inhomogenous term in the Stokes equation we can re-express it as a nonlinear integral equation

$$v_i(x, t) = \int K(x - y \mid t)\, u(y)\, dy$$

$$+ \int_0^t ds \int K(x - y \mid t - s) \left[-\frac{1}{\rho} \partial_i p(y, s) + v_j(y, s) \partial_j v_i(y, s) \right] dy.$$

If the nonlinearities are small enough (i.e., if the initial data is not too big and the time t is not too long) we can solve this equation by iterating it. The zeroth order solution simply ignores the second term:

$$v_i^{(0)}(x, t) = \int K(x - y \mid t)\, u_i(y)\, dy, \quad p^{(0)}(x, t) = 0.$$

The pressure and velocity in each order are determined by solving by putting in the previous order solution into the nonlinear terms:

$$\nabla^2 p^{(n+1)} = \rho \partial_i \left[v_j^{(n)} \partial_j v_i^{(n)} \right] \implies p^{(n+1)}(x) = \int \frac{1}{4\pi |x - y|} \partial_i \left[v_j^{(n)} \partial_j v_i^{(n)}(y) \right] dy,$$

$$v_i^{(n+1)}(x, t) = \int_0^t ds \int K(x - y \mid t - s) \left[-\frac{1}{\rho} \partial_i p^{(n+1)}(y, s) + v_j^{(n)}(y, s) \partial_j v_i^{(n)}(y, s) \right] dy.$$

If the sums

$$\sum_{n=0}^{\infty} v_i^{(n)}(x, t), \quad \sum_n p^{(n)}(x, t)$$

converge, we have solved the NS equations. This method of solving nonlinear differential equations is ubiquitous in physics: the Dyson series of quantum field theory being a more sophisticated example.

12.4   Leray: finite time regularity

Since K and its derivatives are explicitly known, it is possible to estimate the norm of v^{n+1} and p^{n+1} in terms of v^n and p^n. Leray establishes, by a careful analysis, the recursion relation

$$V^{(n+1)}(t) \le A' \int_0^t \frac{\left[V^{(n)}(s) \right]^2}{\sqrt{\nu(t - s)}}\, ds + V^{(n)}(0),$$

where $V^{(n)}(t) = \max_x \sum_{k=0}^{n} v_i^{(k)}(x, t)$ is the maximum speed at time t in the nth order approximation. Also, A' is some constant (i.e., independent of t) depending on the initial data.

If we can find a continuous function $\phi(t)$ satisfying the inequality

$$A' \int_0^t \frac{[\phi(s)]^2}{\sqrt{v(t-s)}} ds + V(0) \le \phi(t),$$

we can conclude by induction that all the $V^{(n)}(t)$ are less than it:

$$V^{(n)}(t) \le \phi(t).$$

Actually a constant $\phi(t) = [1 + A]V(0)$ (for some choice of A) satisfies the integral inequality up to some time

$$\tau = \frac{Av}{V(0)^2}.$$

This establishes the convergence of the perturbation series. For regular initial data, the NS equation has a solution for a finite time interval $0 \le t < \tau$.

We can go a bit further with this argument. Suppose we use the velocity field at time t as initial data. Then it can be continues at least to a new time

$$t + \frac{Av}{V(t)^2}.$$

So if a singularity develops at some T, it must be after this instant:

$$T > t + \frac{Av}{V(t)^2}.$$

Solving for $V(t)$ we get

$$V(t) > \sqrt{\frac{Av}{T-t}}.$$

A singularity might never occur. But if it were to occur, there is some point in space at which the speed diverges like $\frac{1}{\sqrt{T-t}}$.

It is possible to put a finer point on this. Leray shows that the L^p-norms for $p > 3$ would diverge as well: $||v||_p(t) > \frac{1}{(T-t)^{\frac{p-3}{2p}}}$. The condition on maximum speed can be thought of the limiting case $p \to \infty$ of these finer inequalities. But the most interesting case is $p = 3$; recall that this is the scale-invariant norm. In recent work (Seregin, 2015) shows that it must also diverge if a singularity is to form:

$$\limsup_{t \to T^-} ||v(t)||_3 = \infty.$$

12.5   Scale invariant solutions

We can now look at the problem from the opposite angle: try to construct a singular solution. Knowing the rate of divergence at a possible singularity, Leray was led to an ansatz

$$v(x, t) = \sqrt{\frac{a}{T - t}} U\left(\frac{x}{\sqrt{a(T - t)}}\right)$$

for some constant a with the dimensions of kinematic viscosity. To complete the story, we make the ansatz for pressure:

$$p(x, t) = \frac{ab}{T - t}\rho P\left(\frac{X}{\sqrt{a(T - t)}}\right)$$

for some dimensionless constant b. This ansatz is self-similar under the NS scale transformations

$$x \to \lambda x, \quad T - t \to \lambda^2(T - t), \quad v \to \frac{1}{\lambda}v, \quad p \to \frac{1}{\lambda^2}p$$

for any positive λ.

Putting these into the NS equations and choosing $a = 2\nu$, $b = \frac{1}{2}$ gives the time-independent PDE in terms of dimensionless variables:

$$U(x) + x \cdot \nabla U(x) + U \cdot \nabla U = \nabla^2 U - \nabla P,$$
$$\nabla \cdot U = 0.$$

If there is a regular solution to these equations, there would be a *singular* solution to the NS equations. Leray was not able to rule out the existence of such solutions. Recent advances (Necas et al., 1996; Tsai, 1998) show however that such self-similar singularities do not occur.

But perhaps the self-similarity is only discrete. That is, instead of scale invariance under all positive numbers λ, it is only invariant only under some particular value. If we use the ansatz

$$v(x, t) = e^{-\frac{1}{2}\tau} U\left(\tau, xe^{-\frac{1}{2}\tau}\right), \quad \tau = \log\left[a(T - t)\right]$$

we can derive a time-dependent PDE for U. If it has regular periodic solutions, then NS would have singular solutions that are discretely scale invariant. Such singularities do arise in general relativity.

Regularity of NS remains an active frontier in the research on PDEs (Seregin, 2015; Tao, 2016). Solving this problem is the second hardest way to make a million dollars. (The hardest way would be to settle the Riemann hypothesis.)

13

Spectral Methods

Many problems in physics can be reduced to an ordinary differential equation (ODE) in some interval with boundary conditions. Most equations, even linear ones, cannot be solved analytically. We will need to solve them numerically. One approach (Chapter) is to turn the differential equation into a difference equation: the finite difference method. Here, we pursue another approach: to expand the function in some basis and turn the ODE into an algebraic system.

For this purpose, approximating the unknown function as a polynomial would seem to make sense: as the degree of the polynomial grows we should get more accurate approximations. Even so, choosing the monomials x^n as the basis elements is a bad idea. For large n, they point in nearly the same direction in the function space. A much better basis is given by Chebychev polynomials (Aurentz and Trefethen, 2017; Trefethen, 2001).

....

Exercise 13.1 Show that the angle between x^n and x^{n+2} tends to zero as $n \to \infty$ with respect to the inner product $< f, g >= \int_{-1}^{1} f(x)g(x)dx$.

....

We will focus on solving linear equations here. Nonlinear ODEs have to be solved by some iterative method like Newton–Raphson. Each step in this iteration involves solving a linear ODE for which the spectral methods of this chapter can be used. As an application, we will solve a version of the Orr–Sommerfeld equation.

13.1 The Chebychev basis

Chebychev series are to the interval $[-1,1]$ what the Fourier series are to periodic functions. We can use the angular coordinate θ on the interval by setting $x = \cos \theta$. Define

$$T_n(\cos \theta) = \cos n\theta, \quad n = 0, 1, 2 \cdots \tag{13.1}$$

Fluid Mechanics: A Geometrical Point of View. S. G. Rajeev © S. G. Rajeev 2018.
Published in 2018 by Oxford University Press. DOI: 10.1093/oso/9780198805021.001.0001

The functions $T_n(x)$ defined this way are actually polynomials, as can be seen using the recursion relation

$$T_{n+1}(x) = 2xT_n(x) - T_{n-1}(x), \quad n \geq 0.$$

More generally,

$$T_m(x) T_n(x) = \frac{T_{m+n}(x) + T_{n-m}(x)}{2}, \quad m \leq n. \tag{13.2}$$

....

Exercise 13.2 Prove eqn (13.2) using the trigonometric identities $\cos(n \pm m)\theta = \cos n\theta \cos m\theta \mp \sin n\theta \sin m\theta$.

....

The Chebychev polynomials (eqn 13.1) are the orthogonal polynomials w.r.t. the measure $d\theta = \frac{dx}{\sqrt{1-x^2}} \equiv d\mu(x)$:

$$\int_{-1}^{1} T_m(x) T_n(x) d\mu(x) = c_n \delta_{mn}, \quad m, n = 0, 1, 2, \cdots$$

where

$$c_n = \begin{cases} \pi & n = 0 \\ \frac{\pi}{2} & n \neq 0. \end{cases}$$

Thus we can expand a continuous function $f : [-1, 1] \to \mathbb{R}$ in a Chebychev series

$$f(x) = \sum_{n=0}^{\infty} f_n T_n(x),$$

$$f_n = \frac{1}{c_n} \int_{-1}^{1} f(x) T_n(x) dx.$$

Note that

$$T_n(-1) = (-1)^n, \quad T_n(1) = 1.$$

13.2   Spectral discretization

To solve equations numerically, a function $f : [-1, 1] \to \mathbb{R}$ has to be represented by a finite sequence of numbers, for example, its values at a finite sample of points. This information can be repackaged as a finite Chebychev series: the function is approximated by a polynomial with which it agrees exactly at the sample points.

What is a good choice of these sample points? For Fourier series on the circle it is natural to choose equally spaced points in the angular variable, because of the translation invariance of the measure $d\theta$. If we transform this to the coordinate $x = \cos \theta$, we will get

$$x_{j,N} = \cos\left[\frac{\pi}{N}j\right], \quad j = 0, 1, \cdots N.$$

These are the extrema of the Chebyschev polynomials:

$$\frac{dT_N(x)}{dx} = -\frac{d\theta}{dx}N\sin N\theta = N\frac{\sin N\theta}{\sin \theta} = 0 \implies \theta = \frac{\pi}{N}j, \quad 0, 1, \cdots N.$$

The density of these points tends to $d\mu(x)$ as $N \to \infty$: there are more of them near the edges. Solutions to ODEs with boundary conditions tend to vary most rapidly near the boundaries. So it makes sense to place more sample points near the boundary.

13.3 Sampling

Thus, a function is represented as an $N + 1$-dimensional vector f_N with components $f_{j,N} = f(x_{j,N})$. There is a finite Chebyhev series (i.e., a polynomial of order N)

$$f_N(x) = \sum_{k=0}^{N} \tilde{f}_{k,N} T_k(x_{j,N})$$

that agrees with it at the sample points

$$f(x_{j,N}) = \sum_{k=0}^{N} \tilde{f}_{k,N} T_k(x_{j,N}).$$

The transformation $f_{j,N} \rightleftarrows \tilde{f}_{j,N}$ can be accomplished in O $(N \log N)$ operations using the clever fast Fourier transform (adapted to the interval). But we will skip that and use ordinary linear algebra. Define the matrix

$$C_N = \left[T_k(x_{j,N})\right]_{0 \le j,k \le n},$$
$$f_N = C_N \tilde{f}_N.$$

Its inverse will transform in the opposite direction. The polynomial gives an interpolation of f to points not on the original sample.

13.4   Interpolation

Given f_N the value at any x (not necessarily on the grid) can be approximated by $\sum_k P_{k,N}(x)f_{k,N}$, where

$$P_{k,N}(x) = \sum_{j=0}^{N} T_j(x) \left[C_N^{-1} \right]_{jk}.$$

13.5   Differentiation

This interpolation can be used to get an approximation to its derivative, $\sum_k P'_{k,N}(x)f_{k,N}$, where

$$P'_{k,N}(x) = \sum_{j=0}^{N} T'_j(x) \left[C_N^{-1} \right]_{jk}$$

is a polynomial of order $N-1$. We should re-expand it in the Chebychev basis with one fewer sample point to get an $N \times (N+1)$ matrix representing the derivative.

It should be familiar from quantum mechanics that differential operators (momentum, hamiltonian) can be thought of as infinite dimensional matrices. But for our purposes, they are best thought of as *rectangular* matrices rather than square matrices: (Driscoll and Hale, 2015). Each time we differentiate we lose one piece of information about the function: the order of the polynomial interpolant is reduced by one. So we should approximate the derivative by a map from the space of polynomials of order $N+1$ to those or order N, that is, a rectangular matrix of dimension $(N+1) \times (N+2)$. More generally, a differential operator of order r is approximated by a rectangular matrix $(N+1) \times (N+1+r)$. Define the rectangular matrix

$$C'_N = \left[T'_k (x_{j,N-1}) \right]_{0 \le j \le N-1, 0 \le k \le N}.$$

Then the derivative is approximated by

$$D_N = C'_N C_N^{-1}.$$

The operator $\frac{d}{dx}$ has a symmetry we would like to preserve: it changes sign under parity, $x \to -x$. Parity is represented in the discrete version by the flip matrix

$$[\mathcal{J}_N]_{0 \le j,k \le N} = \begin{cases} 1 & j = n - k \\ 0 & \text{otherwise} \end{cases}.$$

For example,

$$\mathcal{J}_3 = \begin{pmatrix} 0 & 0 & 0 & 1 \\ 0 & 0 & 1 & 0 \\ 0 & 1 & 0 & 0 \\ 1 & 0 & 0 & 0 \end{pmatrix}.$$

Then

$$D_N = -\mathcal{J}_{N-1} D_N \mathcal{J}_N,$$

as needed.

13.6 Integration

We look for the discrete version of the definite integral

$$f(x) = \int_0^x g(y)\,dy.$$

To preserve parity we choose the starting of point at the middle of the interval $[-1, 1]$.
It is not difficult to verify that

$$T_k^{(-1)}(x) \equiv \int_{-1}^x T_k(y)\,dy = \begin{cases} \frac{1}{2}\left(\frac{T_{k+1}(x)}{k+1} - \frac{T_{k-1}(x)}{k-1}\right) - \frac{k\sin\left(\frac{\pi k}{2}\right)}{1-k^2}T_0(x) & k \neq 1 \\ \frac{1}{4}\left(T_2(x) + T_0(x)\right) & k = 1 \end{cases}.$$

These are polynomials of order $k + 1$. Define the $(N + 2) \times (N + 1)$ matrices

$$C_N^{(-1)} = \left[T_k^{(-1)}(x_{j,N+1})\right]_{0 \leq j \leq N+1, 0 \leq k \leq N},$$
$$D_N^{(-1)} = C_N^{(-1)} C_N^{-1}.$$

Then we can verify numerically that D_{N+1} is a left inverse of $D_N^{(-1)}$:

$$D_{N+1} D_N^{(-1)} = 1_N. \tag{13.3}$$

By 1_N we mean the $(N + 1) \times (N + 1)$ identity matrix. But it is not a right inverse.
For example,

$$D_3^{(-1)}D_4 = \begin{pmatrix} 1 & 0 & -1 & 0 & 0 \\ 0 & 1 & -1 & 0 & 0 \\ 0 & 0 & 0 & 0 & 0 \\ 0 & 0 & -1 & 1 & 0 \\ 0 & 0 & -1 & 0 & 1 \end{pmatrix} \neq 1_5. \tag{13.4}$$

This matrix has a zero eigenvalue with the constant eigenvector:

$$D_N^{(-1)}D_{N+1} \begin{pmatrix} 1 \\ 1 \\ \cdot \\ \cdot \\ 1 \end{pmatrix} = 0.$$

This must be so as the differentiation kills the constant. Integration also anti-symmetric under the flip:

$$D_N^{(-1)} = -\mathcal{J}_{N+1}D_N^{(-1)}\mathcal{J}_N.$$

We can also check that

$$P_{N+1}(0)D_N^{(-1)} = 0$$

because we chose 0 as the origin of the integration. We are thinking of $P_N(0)$ as a column vector with components $P_{k,N}(0)$.

....

Exercise 13.3 Write a program (e.g., in Mathematica) to calculate the matrices $C_N, D_N, D_N^{(-1)}$ numerically. Verify eqn (13.3) for $N = 3$.

....

13.7   The basic ODE

Let us learn to walk before we run. The discrete version of the basic ODE

$$f'(x) = g(x), \quad f(x_1) = \alpha$$

is

$$D_N f_N = g_{N-1}, \quad P_N(x_1) \cdot f_N = \alpha.$$

Together this can be written as

$$\left(\begin{array}{c} D_N \\ P_N(x_1) \end{array} \right) f_N = \left(\begin{array}{c} g_{N-1} \\ \alpha \end{array} \right).$$

The solution is

$$f_N = D_{N-1}^{(-1)} g_{N-1} + \alpha c_N, \quad c_N = \left(\begin{array}{c} 1 \\ 1 \\ \cdot \\ \cdot \\ \cdot \\ 1 \end{array} \right).$$

Thus, $\left(\begin{array}{c} D_N \\ P_N(x_1) \end{array} \right)$ is a square, invertible matrix of dimension $N+1$. We can numerically invert this square matrix to solve the ODE, without directly using $D_N^{(-1)}$.

This confirms the idea that linear differential operators should be thought of as rectangular matrices. By adding boundary conditions as rows, they become square matrices. By inverting these square matrices we can solve linear ODEs.

Now we can consider more complicated cases.

13.8 Down sampling

This brings up a tricky point. A differential operator of order m could have terms that are lower order. They should all be represented by rectangular matrices of co-dimension m to preserve consistency. Even multiplication by one has to be approximated by a rectangular matrix:

$$I_{m,N} = \left[P_N \left(x_{k,N-m} \right) \right]_{0 \le k \le N-m}.$$

Each row is a vector of dimension $N + 1$ and there are $N - m + 1$ such rows.

Multiplication by a function $V(x)$ is similarly represented by a rectangular (instead of a square diagonal) matrix $I_{m,N} \mathrm{diag}(V_N)$.

....

Exercise 13.4 Solve the driven damped simple harmonic oscillator

$$u'' + \gamma u' + \omega^2 u = e^{8x} \sin \Omega x$$

with the b.c. $u(0) = 2.3, u'(0) = 1.1$ in the range $-1 \le x \le 1$ with $\gamma = 3.0$, $\omega = 12.0, \Omega = 15.1$.

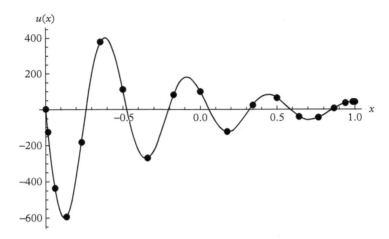

Figure 13.1 *The solution to the damped driven simple harmonic oscillator in the spectral method with* $N = 18$

Solution The discrete version of the equation is

$$\begin{bmatrix} D_{N-1}D_N + \gamma I_{1,N-1}D_N + \omega^2 I_{2,N} \\ P_N(0) \\ P_{N-1}(0)D_N \end{bmatrix} u_N = \begin{bmatrix} f_{N-2} \\ \alpha \\ \beta \end{bmatrix},$$

where $f(x) = e^{8x} \sin \Omega x, \alpha = 2.3, \beta = 1.1$. The solution in the spectral approximation is $u(x) = P_N(x).u_N$. For $N = 18$ the solution is plotted in Figure 13.1.

Example 13.1 Standing waves

Let us solve the equation

$$u'' = \lambda u, \quad u(-1) = 0 = u(1).$$

The eigenvalues are $-\frac{\pi^2}{4}n^2, n = 1, 2, \cdots$. If we discretize as before we will get

$$D_{N-1}D_N u_N = \lambda I_{2,N} u_N,$$
$$P_N(1) \cdot u = 0 = P_N(-1).u,$$

We can solve the boundary conditions by finding an $(N + 1) \times (N - 1)$ matrix Π whose columns are linearly independent vectors satisfying

$$P(1) \cdot \Pi = 0 = P(-1) \cdot \Pi,$$

For example, the NullSpace command in Mathematica gives this matrix Π. (Minor point: you have to take a transpose to get the right shape.)

Then we get the generalized eigenvalue problem

$$D_{N-1}D_N \Pi v_{N-2} = \lambda I_{2,N} \Pi v_{N-2}.$$

We get for $N = 15$,

$$- 2.4674, -9.8696, -22.2066, -39.4784, -61.6884, -88.7312, -122.857, -152.891,$$
$$- 260.471, -285.245, -1123.65, -1188.1, \cdots$$

compared to the exact answers

$$-2.4674, -9.8696, -22.2066, -39.4784, -61.685, -88.8264, -120.903, -157.914,$$
$$-199.859 \cdots$$

If instead we have the b.c.

$$u(-1) = 0 = u'(1),$$

we have the discrete version

$$P_N(-1)u = 0 = P_{N-1}(1)D_N u.$$

The exact answers are

$$-0.61685, -5.55165, -15.4213, -30.2257, -49.9649, -74.6389, -104.248, -138.791,$$
$$-178.27 \cdots$$

to be compared to (for $N = 15$) the discrete approximation:

$$- 0.61685, -5.55165, -15.4213, -30.2257, -49.9649, -74.6425, -104.321, -139.391,$$
$$- 185.615, -273.223, -485.26, -1155.84, -5039.52, \cdots$$

The first $\frac{N}{2}$ eigenvalues are reproduced correctly.

13.9 Spectral solution of the Orr–Sommerfeld equation

We will use a simplified version to make comparisons with (Trefethen, 2001) easier:

$$\frac{1}{\mathcal{R}} \left[\frac{d^4 u}{dx^4} - 2\frac{d^2}{dx^2} + 1 \right] u - 2iu - i(1-x^2)\left[\frac{d^2}{dx^2} - 1 \right] u = \lambda \left[\frac{d^2}{dx^2} - 1 \right] u,$$
$$u(\pm 1) = 0 = u'(\pm 1).$$

The discrete versions of the l.h.s. and r.h.s. are (where $D_N^{(2)} = D_{N-1}D_N$ etc.)

$$A = \frac{1}{\mathcal{R}} \left[D_N^{(4)} - 2I_{2,N-2}D_N^{(2)} + I_{4,N} \right] + -2iI_{4,N} - iV_{N-4}\left[I_{2,N-2}D_N^{(2)} - I_{4,N} \right],$$
$$B = I_{2,N-2}D_N^{(2)} - I_{4,N}.$$

The discrete b.c. are

$$P_N(\pm 1)u = 0 = P_{N-1}(\pm 1)D_N u.$$

Thus Π spans the null space of these four matrices:

$$L = A\Pi, \quad M = B\Pi.$$

Thus λ is a generalized eigenvalue of the pair (L, M). The rightmost eigenvalue in the complex λ-plane is the most unstable one.

Set $\mathcal{R} = 5772$, the critical value of (Orszag, 1971). With $N = 40, 60, 80, 100$ we get the eigenvalue with the smallest real part to be

$$-0.0000780377 - 0.261569i, \quad -0.0000781915 - 0.261568i,$$
$$-0.0000781911 - 0.261568i, \quad -0.0000781903 - 0.261568i.$$

All the other eigenvalues are more stable than this.

Figure 13.2 plots the whole spectrum with $N = 90$. This agrees with the results in (Trefethen, 2001), although our approach is adopted from (Aurentz and Trefethen, 2017). The critical value of $\mathcal{R}$ at which the instability appears is 5812, a bit different than that in (Orszag, 1971). Since the Orr–Sommerfeld operator is not normal, there are transients. The actual instability will kick in before this prediction of the linear theory. So, no need to quibble over this small change.

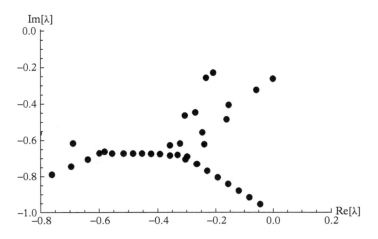

Figure 13.2 *The spectrum of the Orr–Sommerfeld operator with $N = 90$*

13.10   Higher dimensions

We studied in Section 10.6 an example of Fourier spectral methods applied to two-dimensional Euler equations. More realistic than the torus is two-dimensional fluid flow on the sphere, where a similar method based on spherical harmonics can be developed. Chebychev methods do exist in higher dimensions (squares or cubes) but seem to become more cumbersome. The number of nodes increase exponentially with the dimension.

In the finite difference method, the differentiation matrices are sparse. This allows for a larger number of nodes and higher dimensions. Convergence to the continuum solution is slower than in the spectral method. The radial basis function method of Section 14.10 looks promising as a general purpose discretization method, combining advantages of spectral and finite difference methods.

These ideas are important not only for numerical solutions, but for the mathematical foundation of the subject. Once we allow for random forces (most turbulence models have them) partial differential equations (PDEs) are replaced by stochastic differential equations. Solutions are not smooth anymore. The correct analytical formulation is not known yet. A stochastic PDE can also be viewed as a quantum field theory. Renormalization methods needed to remove divergences (singularities) do not yet have a secure mathematical foundation.

In effect, what we seek is an integral calculus of an infinite number variables (functional integration over fields). The modern definition of an integral over a real variable (Lebesgue) was inspired by the numerical methods of the early twentieth century (Burk, 2007). The correct definition of a functional integral must be based on successful numerical methods of our own time.

14

Finite Difference Methods

No human activity has been more transformed by computers than fluid mechanics. There is a vast literature on computational fluid mechanics for the reader wishing to learn that field. This book is about the more geometrical and physical questions. We only take a peek at numerical methods.

We will restrict ourselves to the finite difference method (FDM), the most intuitive of the discrete methods. Finite volume methods are appealing to a physicist because they are based on conservaton laws (Leveque, 2002; Patankar, 1980). The finite element method (FEM) is more versatile, for example, when the boundary has a more complicated shape. Sophisticated connections have been made to discrete methods in topology (Arnold et al., 2010). All this lies outside the scope of this book.

14.1 Differential and difference operators in one dimension

Everyone knows

$$u'(x) = \frac{u(x+h) - u(x)}{h} + \mathrm{O}(h), \tag{14.1}$$

which yields the well-known first order Euler method for solving ordinary differential equations (ODEs). Its iteration is

$$u''(x) = \frac{u(x+h) - 2u(x) + u(x-h)}{h^2} + \mathrm{O}(h^2). \tag{14.2}$$

Less obvious are higher order approximations such as

$$u(x+h) = u(x) + \frac{h}{12}\left[23u'(x) - 16u'(x-h) + 5u'(x-2h)\right] + \mathrm{O}(h^4), \tag{14.3}$$

giving the "third order Adams–Bashforth method" scheme.

Fluid Mechanics: A Geometrical Point of View. S. G. Rajeev © S. G. Rajeev 2018.
Published in 2018 by Oxford University Press. DOI: 10.1093/oso/9780198805021.001.0001

Exercise 14.1 Verify eqn (14.3) by expanding the two sides of the equation in a Taylor series in h.

There is a general method (Fornberg, 2017) to derive the weights $\frac{23}{12}, -\frac{16}{12}, \frac{5}{12}$ in this formula. We illustrate it in this example.

Define the translation operator

$$Tu(x) = u(x+h).$$

The content of Taylor's theorem for analytic functions

$$u(x+h) = u(x) + \frac{h}{1!}u'(x) + \frac{h^2}{2!}u''(x) + \cdots$$

is that the translation operator is the exponential of the differentiation operator:

$$T = e^{h\mathcal{D}}, \quad \mathcal{D} \equiv \frac{d}{dx}.$$

Thus

$$\mathcal{D} = \frac{\log T}{h}.$$

We need to find the weights v_0, w_0, w_{-1}, w_{-2} in

$$u(x+h) = v_0 u(x) + w_0 u'(x) + w_{-1} u'(x-h) + w_{-2} u'(x-2h) + \mathrm{O}(h^4).$$

So we seek the operator identity

$$T = v_0 + \left[w_0 + w_{-1}T^{-1} + w_{-2}T^{-2} \right] \frac{\log T}{h} + \mathrm{O}(h^4).$$

Such manipulations of operators are called "symbol calculus;" they can justified by the Fourier transform, which converts $\mathcal{D}$ and T into multiplication operators.

Continuing in this way,

$$\frac{T - v_0}{w_0 + w_{-1}T^{-1} + w_{-2}T^{-2}} = \frac{\log T}{h} + \mathrm{O}(h^4).$$

Multiplying both sides by T^{-2} and rearranging,

$$T^{-2}\frac{\log T}{h} = \frac{T - v_0}{w_0 T^2 + w_{-1} T + w_{-2}} + \mathrm{O}(h^4).$$

Thus, we are essentially approximating $T^{-2}\frac{\log T}{h}$ by some rational function of T.

14.2   Padé approximant

Many functions with singularities (e.g., branch-cut and double pole as in $z^{-2}\log z$) are not well approximated by a polynomial, that is, a truncated Taylor series is often a poor approximation even around a non-singular point. There is a unique rational function $\frac{P_n(z)}{Q_d(z)}$ (a polynomial of order n divided by a polynomial of order d) such that the Taylor series expansions of $Q_d(z)f(z)$ and $P_n(z)$ around some analytic point z_0 agree up to order $n + d$. This $\frac{P_n(z)}{Q_d(z)}$ is often a good approximation to $f(z)$ near z_0. The whole process can be automated (e.g., in Mathematica). Making the best choice of z_0, n, d is an art: some physical insight is necessary.

With $n = 1, d = 2$, the Padé approximant of $z^{-2}\frac{\log z}{h}$ around $z = 1$ is

$$z^{-2}\frac{\log z}{h} \approx \frac{z - 1}{h\left(\frac{23}{12}(z - 1)^2 + \frac{5(z-1)}{2} + 1\right)}.$$

By extracting the coefficients of powers of z in the numerator and denominator we get the weights

$$v_0 = -1, \quad w_0 = \frac{23}{12}h, \quad w_{-1} = -\frac{4}{3}h, \quad w_{-2} = \frac{5}{12}h.$$

In Mathematica notation,

$$\mathrm{appr} = \mathrm{PadeApproximant}\left[z^{-2}\frac{\mathrm{Log}[z]}{h}, \{z, 1, \{1, 2\}\}\right],$$

$$\mathrm{CoefficientList}[\{\mathrm{Denominator}[\mathrm{appr}], \mathrm{Numerator}[\mathrm{appr}]\}, z],$$

yielding

$$\left\{\left\{\frac{5h}{12}, -\frac{4}{3}h, \frac{23h}{12}\right\}, \{-1, 1\}\right\}.$$

More generally (Fornberg, 2017), the Padé approximant of $z^s\left(\frac{\log z}{h}\right)^m$ of order (n, d) around $z = 1$ will yield various finite difference schemes for solving ODEs. The example above has $s = -2, d = 2, n = 1, m = 1$. Some improvements to the Padé schemes are found in (Fornberg, 2017).

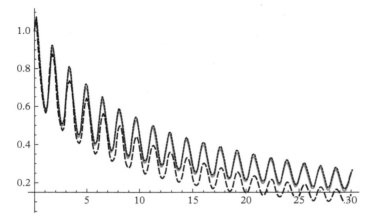

Figure 14.1 *The solid curve is the result of NDSolve in Mathematica. The dashed curve is the solution by first order Euler method. The dotted curve is the third order Adams–Bashforth method, for the same step size* $h = 0.1$

Exercise 14.2 Solve the equation $u'(x) = g(x, u(x))$, where $g(x, u) = x \cos(4x + u)$ with $u(0) = 1$ over the range $0 \leq x \leq 30$. Use the first order Euler method (eqn 14.1) with $h = 0.1$. Compare with the result of NDSolve in Mathematica. Repeat with third order Adams–Bashforth (eqn 14.3). Note the improvement in Figure 14.1 without changing step size h.

14.3   Boundary value problems

As we saw earlier, discrete approximations to differential operators are best thought of as rectangular matrices. Let Ω be some region with boundary $\partial \Omega$. Then the discrete approximation to a second order differential operator (e.g. Laplacian),

$$(L\phi)(x) = \sum_{x' \in \Omega} L(x, x')\phi(x'), \quad x \in \underline{\Omega},$$

maps a function on Ω to functions on its interior $\underline{\Omega} = \Omega - \partial \Omega$. In an inhomogenous equation

$$L\phi = h^2 \rho,$$

the source $\rho(x)$ is a function on $\underline{\Omega}$. (We will find it convenient to think of the discrete approximation L as a dimensionless matrix. $h^2 \rho$ and ϕ have the same dimension.)

The restriction of ϕ to the boundary (for Dirichlet), or its normal derivative (for Neumann) is a linear operator B that takes functions on Ω to those on $\partial \Omega$:

$$[B\phi](x) = b(x), \quad x \in \partial \Omega.$$

Together L, B form a square matrix M acting on the space of functions of Ω. $h^2\rho$ and b together form a function j on Ω. Thus

$$M\phi = j$$

can be solved as a linear system.

For example, if $\Omega = [-1, 1]$ is approximated by n points, $\partial\Omega = \{-1, 1\}$ has two points. Then, L is $(n-2) \times n$ and B is $2 \times n$. The source $\rho(x)$ is an $n-2$-dimensional vector. The discretization of the boundary value problem

$$\phi'' = \rho, \quad \phi(-1) = b_1, \quad \phi(1) = b_2$$

with $n = 6$ gives

$$L = \begin{pmatrix} 1 & -2 & 1 & 0 & 0 & 0 \\ 0 & 1 & -2 & 1 & 0 & 0 \\ 0 & 0 & 1 & -2 & 1 & 0 \\ 0 & 0 & 0 & 1 & -2 & 1 \end{pmatrix}, \quad B = \begin{pmatrix} 1 & 0 & 0 & 0 & 0 & 0 \\ 0 & 0 & 0 & 0 & 0 & 1 \end{pmatrix}.$$

Inserting the first row of B on top of L and the second row of B at the bottom (because the left boundary is the first point and the right boundary the last point) we get the linear equation

$$M\phi \equiv \begin{pmatrix} 1 & 0 & 0 & 0 & 0 & 0 \\ 1 & -2 & 1 & 0 & 0 & 0 \\ 0 & 1 & -2 & 1 & 0 & 0 \\ 0 & 0 & 1 & -2 & 1 & 0 \\ 0 & 0 & 0 & 1 & -2 & 1 \\ 0 & 0 & 0 & 0 & 0 & 1 \end{pmatrix} \phi = \begin{pmatrix} b_1 \\ h^2\rho(x_1) \\ h^2\rho(x_2) \\ h^2\rho(x_3) \\ h^2\rho(x_4) \\ b_2 \end{pmatrix},$$

which can be solved as usual. Of course, it is not necessary to place the rows of B as above; you can place them anywhere without changing the result. It just makes it easier to extract the position dependence of the solution.

....

Exercise 14.3 Solve $\phi'' = 32 \sin(6\pi x) + 3x$, $\phi(\pm 1) = \pm 1$ numerically and compare with the exact solution (Figure 14.2).

....

14.4   Explicit scheme for the diffusion equation

We can apply the FDM to the initial value problem of the diffusion equation (also called the heat equation)

$$\frac{\partial\phi}{\partial t} = \sigma \frac{\partial^2\phi}{\partial x^2}.$$

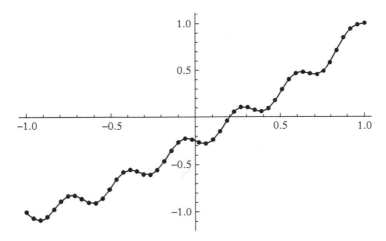

Figure 14.2 *The solid line is the exact solution. The dotted curve is the numerical solution of the boundary value problem*

The simplest approximation by difference operators is

$$\frac{\phi(t+\tau,x)-\phi(t,x)}{\tau}+O(\tau^1)=\sigma\frac{\phi(t,x+h)-2\phi(t,x)+\phi(t,x-h)}{h^2}+O(h^2),$$

yielding the analogue of the Euler scheme:

$$\phi(t+\tau,x)=\phi(t,x)+\frac{\tau}{h^2}\sigma\,[\phi(t,x+h)-2\phi(t,x)+\phi(t,x-h)].\qquad(14.4)$$

The spatial coordinate, x, takes discrete values, spaced in steps of size h.

We can represent the instantaneous value of $\phi(t,x)$ as a vector Φ of size $N+1$ where $h=\frac{2}{N}$. The operator in the square brackets above can now be represented by a tridiagonal matrix

$$L=\begin{pmatrix} -2 & 1 & 0 & . & . & . & 0 \\ 1 & -2 & 1 & . & . & . & 0 \\ 0 & 1 & -2 & 1 & 0 & . & 0 \\ . & 0 & 1 & -2 & 1 & 0 & 0 \\ . & . & 0 & 1 & -2 & 1 & 0 \\ . & . & 0 & 0 & 1 & -2 & 1 \\ 0 & . & . & . & 0 & 1 & -2 \end{pmatrix}.$$

Then the step forward in time can be expressed by a matrix multiplication:

$$\Phi(t+\tau)=\Phi(t)+\frac{\tau\sigma}{h^2}L\Phi(t).\qquad(14.5)$$

It is possible to take advantage of the sparse (tridiagonal) nature of L to speed up this computation.

14.5   Numerical stability

The dimensionless ratio $\frac{\tau\sigma}{h^2}$ determines the stability of this scheme; it is called the *Courant number*. To understand stability, it suffices to consider the case where the spatial dependence is exponential (the general case can be expanded in this basis):

$$\Phi(t, x) = A(t)e^{ikx},$$

$$\Phi(t, x+h) - 2\Phi(t, x) + \Phi(t, x-h) = A(t)2\left[\cos kh - 1\right]e^{ikx} = -4\sin^2\frac{kh}{2}A(t)e^{ikx},$$

$$A(t+\tau) = \left\{1 - 4\frac{\tau\sigma}{h^2}\sin^2\frac{kh}{2}\right\}A(t).$$

The heat equation is dissipative: the magnitude of the solution should not increase in time. So we must have $\mid 1 - 4\frac{\tau\sigma}{h^2}\sin^2\frac{kh}{2} \mid < 1$.

Since $0 \le \sin^2 \le 1$ this becomes

$$\frac{\tau\sigma}{h^2} < \frac{1}{2}. \tag{14.6}$$

The time step should not be too large for a given spatial step. More generally if the diffusion constant is replaced by a variable $\sigma(t, x)$ we would have to make sure that $\frac{\tau\sigma(t,x)}{h^2} < \frac{1}{2}$ over the whole range of values of t, x.

For hyperbolic equations, there is a similar Courant condition on the ratio $\frac{\tau c}{h}$, where c is the speed of propagation. The study of stability is central to numerical analysis and some of the best works are devoted to it (Richtmyer and Morton, 1957).

....

Exercise 14.4 Solve the heat equation in the interval $-1 < x < 1$ with the initial condition $\phi(x) = \sin\pi x$ and boundary conditions $\phi(-1) = 0$, $\phi(1) = 0$. Use the explicit scheme above. Compare with the exact solution $\phi(t, x) = e^{-\pi^2 t}\sin\pi x$. Check that $\tau > \frac{1}{2}h^2$ leads to an instability.

....

14.6   Implicit schemes

A way around this instability is to replace the correction term on the r.h.s. of eqn (14.5) by a weighted average of its values at t and $t+\tau$. Then we would have to solve for $\Phi(t+\tau)$, instead of simply computing it from $\Phi(t)$: hence the name *implicit scheme*. Even for ODEs when there are vastly different rates of time dependence ("stiff" equations), we will need such alternatives to the explicit schemes (Iserles, 1996).

$$\Phi(t + \tau) = \Phi(t) + \frac{\tau\sigma}{h^2} L\left[(1 - \theta)\Phi(t) + \theta\Phi(t + \tau)\right], \quad 0 \le \theta \le 1.$$

The same analysis as above gives

$$A(t + \tau) = A(t) - 4\frac{\tau\sigma}{h^2}\sin^2\frac{kh}{2}\left[(1 - \theta)A(t) + \theta A(t + \tau)\right].$$

The ansatz $A(t + \tau) = \lambda A(t)$ gives the rate of growth

$$\lambda = \frac{4\alpha\theta - 4\alpha + 1}{4\alpha\theta + 1}, \quad \alpha = \frac{\tau\sigma}{h^2}\sin^2\frac{kh}{2}.$$

The stable region $-1 \le \lambda \le 1$ (plotted in Figure 14.3) is

$$\frac{\tau\sigma}{h^2} \le \frac{1}{2(1 - 2\theta)}, \quad 0 \le \theta < \frac{1}{2}.$$

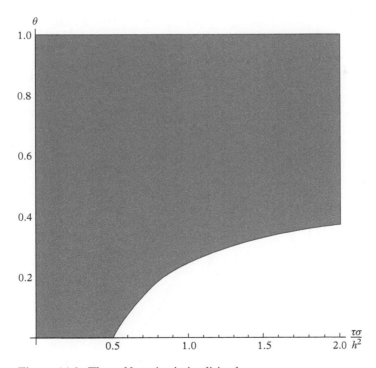

Figure 14.3 *The stable region in implicit schemes*

If $\theta \geq \frac{1}{2}$ it is stable for all Courant numbers. The discrete Laplacian being sparse (tridiagonal in this case, band diagonal in higher dimensions) lightens the load of having to solve for $U(t+\tau)$. Such linear equations can be solved in O(N) steps instead of O(N^3) (or O($N^2 \log N$) if you use memory-intensive methods) for generic linear equations.

14.7   Physical explanation

Why are implicit schemes stable while explicit schemes can be unstable? Let us imagine what happens in between two time steps. The diffusion equation is dissipative, so each of the modes (eigenfunctions of the Laplacian) decay exponentially. So the right interpolation to intermediate times is a weighted exponential (not a polynomial as we pretend when we do expansions in h, τ) in time. Because exponential decay is so fast, the value at the later time is much closer to the correct value than at the earlier time. This suggests that the choice $\theta = 1$ (totally omitting the earlier time) is best. This is indeed what is argued in engineering texts (Patankar, 1980).

....

Exercise 14.5 Solve the heat equation in the interval $-1 < x < 1$ with the initial condition $\phi(x) = \sin \pi x + 3 \sin 2\pi x$ and boundary conditions $\phi(-1) = 0$, $\phi(1) = 0$.

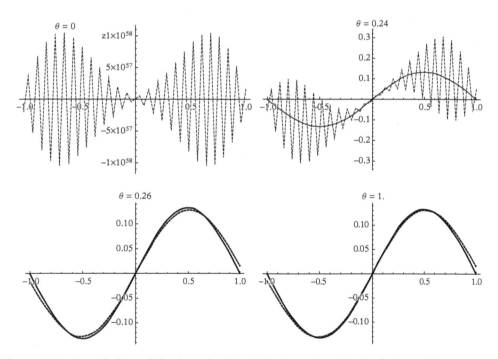

Figure 14.4 $\theta = 0$ *is the explicit scheme,* $\theta = 0.24$ *is just below the critical value for stability,* $\theta = 0.26$ *just above it and* $\theta = 1.0$ *is the fully implicit scheme. The solid line is the exact solution*

Choose Courant number $\frac{\tau\sigma}{h^2} = 1$, $N = 50$ and compare after 128 time steps for $\theta = 0, 0.24, 0.26, 1.0$.

Solution See Figure 14.4.

....

14.8   The Poisson equation

The Poisson equation $\nabla^2\phi = \rho$ is a staple of mathematical physics. It is most familiar in electro-statics, but was first recognized (in its homogenous form, the Laplace equation) in ideal incompressible fluid flows. The most obvious approximation is to use eqn (14.2) in each direction. In two dimensions,

$$\nabla^2\phi = \frac{1}{h^2}\left[\phi(x+h,y) + \phi(x-h,y) + \phi(x,y+h) + \phi(x,y-h) - 4\phi(x,y)\right] + O(h^2).$$

We can understand this in terms of the translation operators

$$T_1 = e^{h\partial_1}, \quad T_2 = e^{h\partial_2},$$

$$\nabla^2 = \left(\frac{\log T_1}{h}\right)^2 + \left(\frac{\log T_2}{h}\right)^2.$$

The above approximation amounts to

$$\left(\frac{\log T_1}{h}\right)^2 + \left(\frac{\log T_2}{h}\right)^2 - \frac{1}{h^2}\left[T_1 + T_1^{-1} + T_2 + T_2^{-1} - 4\right] = O(h^2).$$

The simplest way to verify this is to make the substitution $T_1 = 1 + hy_1$, $T_2 = 1 + hy_2$ and expand in h around $h = 0$.

Again, we can improve accuracy be using symbol calculus. For example (Fornberg and Flyer, 2001; Iserles, 1996),

$$P(T_1, T_2) = -20 + 4\left(T_1 + T_1^{-1} + T_2 + T_2^{-1}\right) + \left(T_1 + T_1^{-1}\right)(T_2 + T_2^{-1}),$$

$$Q(T_1, T_2) = 8 + T_1 + T_1^{-1} + T_2 + T_2^{-1}.$$

Then, the above substitution and expansion gives

$$\left(\frac{\log T_1}{h}\right)^2 + \left(\frac{\log T_2}{h}\right)^2 - \frac{2}{h^2}\frac{P(T_1, T_2)}{Q(T_1, T_2)} = O(h^4).$$

This means that the Poisson equation can be approximated by

$$\frac{1}{h^2}P(T_1, T_2)\phi = \frac{1}{2}Q(T_1, T_2)\rho + O(h^4).$$

This amounts to including next-to-nearest (diagonal) terms (Fornberg and Flyer, 2001; Iserles, 1996); in the stencil notation favored by numericists,

$$\frac{1}{h^2}\begin{bmatrix} 1 & 4 & 1 \\ 4 & -20 & 4 \\ 1 & 4 & 1 \end{bmatrix}\phi = \frac{1}{2}\begin{bmatrix} & 1 & \\ 1 & 8 & 1 \\ & 1 & \end{bmatrix}\rho + O(h^4). \tag{14.7}$$

It might be useful to develop a multivariate Padé algorithm producing such formulas, generalizing (Fornberg, 2017).

A way to solve the Laplace equation, starting with an initial guess ϕ_0, is to solve the heat equation with ϕ_0 as the initial condition. As $t \to \infty$ the heat solution will tend to the Laplace solution. Any function ϕ_0 satisfying the boundary conditions can be the starting point. Adding a source term to the heat equation, we can also apply this strategy to the Poisson equation.

....

Exercise 14.6 Solve the Poisson equation in the plane with $\rho(x,y) = -15\pi^2 \sin(\pi x) \sin(2\pi y)$ the boundary conditions $\phi(\pm1,y) = \pm\sinh(\pi)\sin(\pi y)$, $\phi(x,\pm1) = 0$. Use the five point discretization (eqn 14.7) above. Compare with the exact solution $\phi(x,y) = 3\sin(\pi x)\sin(2\pi y) + \sinh(\pi x)\sin(\pi y)$.

Solution See Figure 14.5.

....

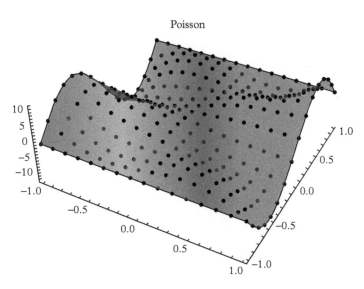

Figure 14.5 *Exact solution of a Poisson equation compared to the numerical solution*

....

Exercise 14.7 Numerically solve the static two-dimensional Navier–Stokes equations in a square box with the boundary conditions that the velocity is zero on three sides, but is a non-zero constant on one side (i.e., one wall is sliding at a constant velocity). The discretized nonlinear equations can be solved using the Newton–Raphson method, for example, FindRoot in Mathematica (Mokhasi, 2013).

....

14.9 Discrete version of the Clebsch formulation

The velocity of an incompressible fluid can be parametrized by the Clebsch variables α, β:

$$v = \nabla\gamma + \frac{1}{2}\left[\alpha\nabla\beta - \beta\nabla\alpha\right].$$

γ is determined as a function of α, β by solving the Poisson equation

$$\nabla \cdot v = 0 \iff \nabla^2\gamma = -\frac{1}{2}\nabla \cdot [\alpha\nabla\beta - \beta\nabla\alpha].$$

Euler equations (in vorticity form) are equivalent to the advection equations for these scalars:

$$\frac{\partial\alpha}{\partial t} + v \cdot \nabla\alpha = 0, \quad \frac{\partial\beta}{\partial t} + v \cdot \nabla\beta = 0.$$

Once v is eliminated, these are nonlinear evolution equations for α, β.

Although rarely used, a discrete version of these equations is quite natural. It would be interesting to see how well these represent flows of interest in engineering and physics.

Think of the discretization of space as a graph: a finite set V of points (vertices) and a subset E of pairs $(x, y) \subset V \times V$ as edges. The graph is symmetric: $(x, y) \in E \implies (y, x) \in E$ as well. Each edge has a length $|x - y|$: not all of them have to be equal. This gives us the freedom to chose a finer mesh near a region where the fields are likely to vary more rapidly (e.g., near the edges).

Scalar fields reside at the vertices: $\alpha : V \to \mathbb{R}$ etc. The derivative of scalar lives on the edges

$$\nabla\alpha : E \to \mathbb{R}, \quad \nabla\alpha(x, y) = \frac{\alpha(x) - \alpha(y)}{|x - y|}, \quad (x, y) \in E.$$

A vector field (such as v) is also a function from edges to the reals, anti-symmetric under the interchange of end-points:

$$v(x, y) = -v(y, x).$$

The discrete version of the Clebsch parametrization is

$$
v(x,y) = \frac{1}{|x-y|} \left\{ \gamma(x) - \gamma(y) + \frac{1}{2} \left[\alpha(x) \{\beta(x) - \beta(y)\} - \beta(x) \{\alpha(x) - \alpha(y)\} \right] \right\}.
$$

That is,

$$
v(x,y) = \frac{1}{|x-y|} \left\{ \gamma(x) - \gamma(y) + \frac{1}{2} [\beta(x)\alpha(y) - \alpha(x)\beta(y)] \right\}.
$$

Since the density is constant, $v(x,y)$ can be thought of as the mass that flows from y to x. Incompressibility is the conservation law that the total flux into a point must be zero. Let $N(x)$ be the set of neighbors of x:

$$
N(x) = \{y | (x,y) \in E\}.
$$

Then incompressibility is the condition

$$
\sum_{y \in N(x)} v(x,y) = 0.
$$

This leads to an algorithm for solving the discrete version of Euler's equation. In simplified form:

1. Given α, β the at time t, solve the discrete Poisson equation

$$
\sum_{y \in N(x)} \frac{\gamma(t|x) - \gamma(t|y)}{|x-y|} = \frac{1}{2} \sum_{y} \frac{\alpha(t|x)\beta(t|y) - \beta(t|y)\alpha(t|x)}{|x-y|}
$$

to get $\gamma(t,x)$. The l.h.s. is a sparse matrix, as E is a very small subset of $V \times V$.

2. Then

$$
v(t|x,y) = \frac{1}{|x-y|} \left\{ \gamma(t|x) - \gamma(t|y) + \frac{1}{2} [\beta(t|x)\alpha(t|y) - \alpha(t|x)\beta(t|y)] \right\}
$$

gives the velocity at time t.

3. The implicit equations

$$
\alpha(t+\tau|x) + \tau \sum_{y \in N(x)} v(t|x,y) \frac{\alpha(t+\tau|x) - \alpha(t+\tau|y)}{|x-y|} = \alpha(t|x),
$$

$$
\beta(t+\tau|x) + \tau \sum_{y \in N(x)} v(t|x,y) \frac{\beta(t+\tau|x) - \beta(t+\tau|y)}{|x-y|} = \beta(t|x)
$$

determine the Clebsch scalars at the next instant. Again, this involves inverting a sparse matrix.

To complete this story, we would like to have a Clebsch formalism for the Navier–Stokes equation. Without dissipation, one worries about the stability of this scheme, in spite of its implicit nature.

14.10 Radial basis functions

All finite methods (FDM, FEM, spectral methods, etc.) are ways of representing a field by a finite list of values. These could be values of the field at some finite set of points ("nodes"). Then the values at other points in space have to be found by some interpolation. If this interpolation is by piecewise polynomial functions we get FDM or its variants (FEM etc.). If by a Fourier or Chebycheff series we get spectral methods. The latter are very accurate, but seem limited to one dimension: the differentiation matrices are not sparse and are simply too large for n^2 or n^3 variables. The former do work in higher dimensions but are not as accurate. A way of combining the efficacy of the two methods is the radial basis function (RBF) method (Buhmann, 2003; Fornberg and Flyer, 2001; Wendland, 2005).

Given the values of a function $f(q_r)$ at some (pairwise unequal) points q_r we can interpolate (Buhmann, 2003; Fornberg and Flyer, 2001; Wendland, 2005) to an arbitrary point x:

$$f(x) = \sum_r f_r \phi (x - q_r) . \tag{14.8}$$

The coefficients f_r are determined by solving the linear system

$$f(q_s) = \sum_r \phi_{sr} f_r, \quad \phi_{sr} = \phi (q_s - q_r),$$

which guarantees that the interpolant agrees with the given data at q_s. The RBF $\phi(x)$ is rotation invariant. A popular choice is

$$\phi(x) = e^{-|\epsilon x|^2}, \quad \epsilon > 0.$$

But it is not necessary in general that the RBF ϕ be a decreasing function. For example (Fornberg and Flyer, 2001), $\phi(x) = |x|^3$ corresponds to cubic spline interpolation; more generally, $|x|^{2k+1}$ (with an odd power) interpolates with a polynomial spline of order $2k + 1$. The wide gaussian ($\epsilon \to 0$) is equivalent to Lagrange polynomial interpolation with nodes q_r. (The matrix ϕ_{sr} becomes ill-conditioned as $\epsilon \to 0$, but there is a way (Fornberg and Flyer, 2001) to "renormalize" this and find an equivalent basis which is well-conditioned for small ϵ.)

So what is the best value of ϵ? If it is too small, the interpolant will suffer from the Runge phenomenon (analog of the Gibbs phenomenon for polynomial interpolation): it

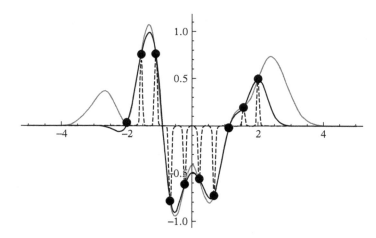

Figure 14.6 *RBF interpolation for $N = 10$ equally spaced nodes q_r in the interval $[-R, R]$ with $R = 2$. The black curve has $\epsilon = \frac{N}{2R}$; the dashed curve has eight times the density and the gray curve has half. When ϵ is too big (dashed curve) the interpolant wants to hug the real axis with jumps at each of the nodes to reach the required value. Notice the Runge oscillations near the boundary for too small a value of ϵ (gray curve)*

will oscillate wildly at the boundary of the range of the nodes q_r. If ϵ is too big, there will be spurious oscillations in between the nodes. The optimal choice is $\epsilon \approx$ the density of nodes. See Figure 14.6.

If the nodes are not equally spaced (e.g., randomly chosen) we must generally pick a larger value of ϵ because of the possibility of occasional near-collisions between nodes. Of relevance to fluid mechanics is the case where the data is a vector v_r at each node q_r. (McNally, 2011) shows how to interpolate it as a divergence-free vector field:

$$v_a(x) = \sum_r \phi_{ab} (x - q_r) v_{rb}, \quad a = 1, 2, 3, \quad r = 1, \cdots N. \tag{14.9}$$

The RBF for vector fields can be derived from that of scalars, in Euclidean space:

$$\phi_{ab}(x) = \left\{ \partial_a \partial_b - \nabla^2 \delta_{ab} \right\} \phi(x). \tag{14.10}$$

On curved manifolds (e.g., surface of the Earth) the generalization of a gaussian is the kernel of the heat equation at "time" ϵ. The heat kernel on vector fields is an independent function: not derivable, in general, from the scalar kernel. It will have the isometries of the underlying manifold (e.g., rotation symmetry in the case of a sphere).

(McNally, 2011) uses a square grid of points q_r and discretizes the magnetohydrodynamics (MHD) equations. The divergence-less vector field of interest is the magnetic field: the fluid is compressible in his application, so the velocity need not have zero divergence.

After expansion in the basis above, the MHD equations reduce to ODEs which can be solved numerically. Many applications to fluid mechanics and geosciences are discussed by (Fornberg and Flyer, 2001).

14.11   A Lagrangian discretization

All of the methods we have mentioned so far have been discretizations of the Eulerian picture of fluid mechanics. We mention here an idea for another discretization that leads to a Lagrangian picture. This needs to be explored further.

Suppose we have approximated an incompressible vector field as above (eqn 14.9). The energy (choosing units with density equal to one) is

$$H = \frac{1}{2} \int v^2(x) dx = \frac{1}{2} g_{ij}(q) v^i v^j,$$

where

$$g_{ij}(q) = \int \phi_{ac} (x - q_r) \phi_{cb} (x - q_s) \, dx.$$

Here, $i \equiv ar$ and $j \equiv bs$. We will denote $q^i \equiv q_{ar}$ and $v^i \equiv v_{ar}$ as contravariant components following the conventions of geometry. We can regard the positive matrix $g_{ij}(q)$ as a Riemannian metric on the configuration space (Cohen, 2010) of "particles" with pairwise unequal positions q_r. This Riemannian manifold can be thought of as a finite dimensional approximation to the group $S\mathcal{D}$ of volume preserving diffeomorphisms, with the L^2-metric providing a (left invariant) Riemannian metric. The equations

$$\frac{dq^i}{dt} = v^i,$$

$$\frac{dv^i}{dt} + \Gamma^i_{jk}(q) v^j v^k = 0$$

then become the discrete approximation to Euler's equations of ideal incompressible fluid mechanics. Here

$$\Gamma^i_{jk} = \frac{1}{2} g^{il} \left[\partial_j g_{lk} + \partial_k g_{lj} - \partial_l g_{jk} \right]$$

are the Christoffel coefficients determined by the metric g. There is a even a natural extension that allows for viscosity.

The norm of the gradient of the velocity field defines another metric on the same configuration space:

$$\int |\nabla v|^2 dx = \gamma_{ij}(q)v^i v^j.$$

This gives a natural dissipation term:

$$\frac{d^2 q^i}{dt^2} + \Gamma^i_{jk}(q)\frac{dq^j}{dt}\frac{dq^k}{dt} = -v g^{ik}(q)\gamma_{kl}(q)\frac{dq^l}{dt}.$$

It would be interesting to see how well these equations approximate fluid flows.

Exercise 14.8 Calculate explicitly the components of the metric tensor g_{ij}, using the Gaussian RBF and eqn (14.10). Compute the curvature of this metric in the simplest case with $N = 2$.

15

Geometric Integrators

Discrete approximations underlying the numerical methods can break the natural symmetries of the original physical or geometrical problem. This leads to problems: loss of conservation laws, numerical instabilities, or stability when the underlying physical system is in fact unstable.

Physically speaking, you would want your numerical integrator to conserve energy (or angular momentum) for conservative systems, and to have the correct dissipation in systems with friction. Because of Noether's theorem in mechanics, symmetries and conservation laws are closely intertwined. So the geometric and physical issues are not separate.

Geometric integrators are clever modifications of the usual numerical methods (Runge-Kutta etc.) for ordinary differential equations (ODEs) which preserve geometric structures: symplectic integrators in mechanics (Rajeev, 2013), Lie group methods when the time evolution is through a group action (Iserles et al., 2000), etc.

Similar geometric integrators for partial differential equations are not yet fully developed. One approach would be to generalize the Lie theoretic methods for ODEs to infinite dimensional groups, for example, in fluid mechanics, to the group of diffeomorphisms. It might lead to a discretization of Euler and Navier–Stokes that preserves the geometry we explored in Chapter 11. In this chapter we will survey some techniques for ODEs to prepare for this direction of research.

15.1 Lie group methods

Suppose e_a is a basis for the Lie algebra $\mathcal{G}$, with structure constants C_{ab}^c:

$$[e_a, e_b] = C_{ab}^c e_c.$$

A representation of $\mathcal{G}$ on some vector space is given by matrices satisfying

$$[\lambda_a, \lambda_b] = C_{ab}^c \lambda_c.$$

Many interesting equations of mathematical physics can be written in the form

$$\frac{dq}{dt} = u(t, q)q(t), \quad u(t, q) = u^a(t, q)\lambda_a, \quad q(0) = q_0 \tag{15.1}$$

Fluid Mechanics: A Geometrical Point of View. S. G. Rajeev © S. G. Rajeev 2018.
Published in 2018 by Oxford University Press. DOI: 10.1093/oso/9780198805021.001.0001

for some functions of "time" t and "position." We saw several examples in Chapter 10 including the rigid body and some fluid models.

As always, we begin the study with the linear equations of this type where $u(t)$ depends on time but not position. In this case the solution can be written as in terms of the group action $q(t) = g(t)q_0$ where $g(t)$ is the solution to

$$\frac{dg}{dt} = u(t)g(t), \quad u(t) = u^a(t)\lambda_a, \quad g(0) = 1.$$

Exquisite methods of Lie theory will give a solution to this equations (in terms an infinite series of multiple commutators). It can be used to get numerical approximation methods for the nonlinear case (eqn 15.1) as well.

15.2   Exponential coordinates

If $u(t)$ happens to be independent of t the solution to eqn (15.1) is simply $g(t) = \exp(tu)$. Here the matrix exponential $\exp : \mathcal{G} \to \mathfrak{G}$ is defined by the infinite series

$$\exp(u) = \sum_{r=0}^{\infty} \frac{1}{r!} u^r.$$

This matrix exponential has an inverse (a "matrix logarithm") within a sufficiently small neighborhood of the identity in $\mathfrak{G}$. So, it provides a coordinate system (called the "canonical coordinate system of the first kind" (Varadarajan, 1984)) in this neighborhood, in which eqn (15.1) can be written more explicitly, allowing numerical approximations to be developed.

....

Exercise 15.1 Use the Pauli basis $e_a = \frac{1}{2i}\sigma_a$ in $su(2)$ to show that the matrix exponential is

$$\exp\left(e_a u^a\right) = \cos\frac{|u|}{2} + \frac{2}{|u|}e_a u^a \sin\frac{|u|}{2},$$

where $|u| = \sqrt{u^a u^a}$.
Hint Show first that $(e_a u^a)^2 = -\frac{1}{4}|u|^2$.
....

There are other useful coordinate systems as well, for example, the Cayley map

$$\mathrm{cay}(u) = \left(1 - \frac{1}{2}u\right)^{-1}\left(1 + \frac{1}{2}u\right)$$

in $SO(n)$. (The order of multiplication does not matter as the two factors commute.)

....

Exercise 15.2 Show that:

1. The Cayley map $u \mapsto \frac{1+\frac{1}{2}u}{1-\frac{1}{2}u}$ maps the imaginary axis to the unit circle.
2. When u is an anti-symmetric matrix, cay(u) is an orthogonal matrix.

....

15.3 Differentiating the matrix exponential

To write eqn (15.1) in exponential coordinates we will need to differentiate $g(t) = \exp [U(t)]$. This is tricky, as $[U(t), U(t')]$ may not vanish for unequal times. A useful formula (Hausner and Schwartz, 1968) is

$$\frac{d}{dt} \exp [U(t)] = \int_0^1 \exp [sU(t)] \frac{dU(t)}{dt} \exp [-sU(t)] \, ds \, \exp [U(t)].$$

A more explicit form is obtained by expanding the integrand in a series and integrating term by term:

$$\frac{d}{dt} \exp [U(t)] = \left\{ \dot{U}(t) + \frac{1}{2!}[U(t), \dot{U}(t)] + \frac{1}{3!}[U(t), [U(t), \dot{U}(t)]] + \cdots \right\} \exp [U(t)].$$

It is useful to write this in a more compact notation:

$$\frac{d}{dt} \exp [U(t)] = \psi \left(\hat{U}(t) \right) \frac{dU(t)}{dt} \exp [U(t)].$$

Here

$$\psi(x) = \int_0^1 e^{sx} ds = \frac{e^x - 1}{x} = 1 + \frac{1}{2!}x + \frac{1}{3!}x^2 + \cdots$$

is an entire function. Also, $\hat{A}$ is the linear operator on $\mathcal{G}$ given by the Lie bracket

$$\hat{A}B = [A, B].$$

(A popular notation is to denote what we call $\hat{A}$ as ad$A$.)

....

Exercise 15.3 Show that $\psi(x)$ is an entire function that vanishes nowhere. Therefore its reciprocal

$$\beta(x) = \frac{x}{e^x - 1}$$

is also an entire function. $\beta(x)$ is the generating function of the Bernoulli numbers B_r:

$$\frac{x}{e^x - 1} = \sum_{r=0}^{\infty} B_r \frac{x^r}{r!}.$$

Calculate the first few terms:

$$\frac{x}{e^x - 1} = 1 - \frac{1}{2}x + \frac{1}{12}x^2 - \frac{1}{720}x^4 + \cdots$$

....

Thus eqn (15.1) becomes

$$\psi\left(\hat{U}(t)\right)\frac{dU(t)}{dt} = u(t) \implies \frac{dU(t)}{dt} = \beta\left(\hat{U}(t)\right)u(t). \tag{15.2}$$

15.4   Magnus expansion

This nonlinear differential equation can now be solved by iteration, to get an infinite series involving a commutator, double commutators, triple commutators and so on. To second order

$$U(t) = \int_0^t u(s)ds - \frac{1}{2}\int_0^t ds_1 \int_0^{s_1} ds_2 \, [u(s_2), u(s_1)] \tag{15.3}$$

$$+ \frac{1}{12}\int_0^t ds_1 \int_0^{s_1} ds_2 \int_0^{s_1} ds_3 \, [u(s_3), [u(s_2), u(s_1)]]$$

$$+ \frac{1}{4}\int_0^t ds_1 \int_0^{s_1} ds_2 \int_0^{s_2} ds_3 \, [[u(s_3), u(s_2)], u(s_1)] + \cdots$$

Each term is determined by the numerical factor in front, the region of integration over the intermideate times $s_1, s_2, \cdots$ and the pattern of nested commutators. There are exquisite diagrammatic rules (Iserles, 2002), in terms of rooted binary trees, that give you all of these ingredients to any order. This expansion of Magnus is reminiscent of the Dyson series and Feynman diagrams in quantum field theory. The Magnus series (eqn 15.3) converges in some norm $|,|$ on the Lie algebra if $\int_0^t |u(s)|ds < 1$.

For numerical work, the integrals can themselves be approximated by a sum (e.g., using the Gauss–Legendre quadrature formula) to the desired accuracy. The theory of free Lie algebras can be used to get formulas with the least number of commutators (which are computationally expensive; (Iserles et al., 2000)).

We will now give a couple of the geometric variants of the conventional methods for solving ODE (15.1).

Example 15.1 Explicit Euler method

This is the most naive method. For times $t_n = n\tau$ that are equally spaced

$$q_{n+1} = \exp\{\tau u(t_n, q_n)\} q_n. \tag{15.4}$$

Example 15.2 Trapezoidal rule

$$F_1 = \tau u(t_n, q_n), \quad q_{n+1} = \exp\left\{\frac{1}{2}(F_1 + F_2)\right\} q_n$$

where F_2 is the solution of the implicit equation

$$F_2 = \tau u\left(t_n + \tau, \exp\left\{\frac{1}{2}(F_1 + F_2)\right\} q_n\right)$$

It is explicit if $u(t, q)$ is independent of q (linear case). This is a non-commutative version of the popular trapezoidal quadrature formula.

Geometric integrators corresponding to most of the usual methods, such as Runge-Kutta, for solving ODEs have been developed. See (Iserles et al., 2000).

Appendix A
Dynamical Systems

Poincaré was the first to advocate the study of dynamical systems as iterations of a function, focusing on qualitative (topological) properties. A more modern source is (Smale, 1967). Like linear algebra or calculus, dynamical systems have become a general purpose tool in every area of physics: even Wilson's theory of renormalization in quantum field theory relies on it. Every physicist needs to be familiar with it.

A smooth function $f : M \to M$ of a manifold to itself, with a smooth inverse, is called a *diffeomorphism*. The *orbit* of a point is the sequence (infinite in both directions)

$$\cdots f^{-1}\left(f^{-1}(x)\right), f^{-1}(x), x, f(x), f(f(x)), f\left(f(f(x))\right) \cdots$$

The number of repetitions of f is a discrete analog of time: we should think of a diffeomorphism f as defining a dynamical system. It is convenient to write $f^k(x)$ for the action of f repeated k times, if $k > 0$. Of course, if $k = 0$, we have the identity map $f^0(x) = x$. If $k < 0$, $f^k(x)$ is the repetition $|k|$ times of f^{-1}.

- A point that is unchanged under the action of f is a *fixed point*: $f(x) = x$.
- If m is the smallest positive number for which $f^m(x) = x$, we say that x is a *periodic point* with period m.

Thus, a periodic point is simply a fixed point of some iterate of f. So we won't have to study them separately.

A.1   Jacobi matrix at a fixed point

Near a fixed point we can expand

$$f(x) = f(p) + f'(p)\,(x - p) + \cdots,$$

where f' is the matrix of derivatives of f evaluated at p. Since f is invertible, f' also is invertible.

The eigenvalues and eigenvectors of this matrix determine the local behavior of f (and of a finite number of its iterations) near the fixed point. The tangent space (the set of all vectors at p) splits a sum $V(p) = V^s(p) \oplus V^c(p) \oplus V^u(p)$ into a *stable subspace* $V^s(p)$ (eigenvalues of magnitude less than one), a *center subspace* $V^c(p)$ (eigenvalues of magnitude one) and an *unstable subspace* $V^u(p)$ (eigenvalues of magnitude greater than one). These subspaces are transversal but not necessarily orthogonal, for example, the vectors in $V^s(p)$ are at some non-zero angle to those in $V^u(p)$ (with respect to an inner product) but not necessarily at right angles.

The fixed point is *hyperbolic* (or is a saddle) if all the eigenvalues of f' are of magnitude not equal to one; that is, if V^c is empty. Such matrices are generic, that is, under a small perturbation, every matrix becomes hyperbolic. (More precisely, the subset of hyperbolic matrices is open and dense within the space of all invertible matrices.) So we restrict our study to the hyperbolic case for now. (Center spaces are also interesting sometimes, e.g., "marginal" perturbations in renormalization theory.)

....

Exercise A.1 Find the stable and unstable sub-spaces of the linear map $(x^1, x^2) \mapsto (2x^1 + x^2, x^1 + x^2)$.

....

It is clear that under iterations of f, points infinitesimally close to p will get even closer to p if they start within V^s; and are driven away outside the region of validity of the linear approximation if there is even a small component in V^u. What happens once we are out of the immediate neighborhood of p cannot be determined within the linear approximation. We now go beyond.

A.2   Stable and unstable manifolds

Consider a diffeomorphism $f : M \to M$ with a hyperbolic fixed point p. The set of points x in M such that $f^k(x) \to p$ as $k \to \infty$ is called the *stable manifold* $W^s(p)$ of p. Similarly, the *unstable manifold* $W^u(p)$ is the set of points such that $f^k(x) \to p$ as $k \to -\infty$ (note that W^s and W^u get interchanged under $f \rightleftarrows f^{-1}$):

$$W^s(p) = \left\{ x \mid \lim_{k \to \infty} f^k(x) = p \right\}, \quad W^u(p) = \left\{ x \mid \lim_{k \to -\infty} f^k(x) = p \right\}.$$

Example A.1

On the Riemann sphere, the map $f(z) = 2z$ has two fixed points $z = 0, \infty$. Then $W^s(0) = \{0\}$, $W^u(0) = \mathbb{S}^2 - \{0\}$, $W^s(\infty) = \mathbb{S}^2 - \{0\}$, $W^u(\infty) = \{\infty\}$

It is clear that $V^s(p)$ is tangential to $W^s(p)$ and $W^u(p)$ to $V^u(p)$. But away from the vicinity of p the manifolds $W^{s,u}(p)$ can have very complicated shapes.

In particular, the stable manifold $W^s(p)$ can fold around and *intersect* the unstable manifold $W^u(p)$. That is, there may be a point h that is attracted to p under both *forward and backward* iteration. When the intersection is transversal (a vector at the intersection h can be split into the sum of one tangential to $W^s(p)$ and another tangential to $W^u(p)$), such points are called *homoclinic points*. Poincare' missed the possibility of such *homoclinic points* at first[1] but then realized that they are central to the phenomenon of chaos.

---

[1] The legend is that Mittag-Leffler, the editor of the journal *Acta Mathematica*, tried to get back every copy of the issue containing Poincarés wrong theorem, to destroy them. But he missed one copy (Barrow-Green, 1996).

Some contemplation shows that

- there is a neighborhood of h that does not contain other homoclinic points (this is the meaning of transversality);
- the forward and backward iterates of a homoclinic point are also homoclinic. So, if there is one homoclinic point, there are an infinite number of them.

....

Exercise A.2 Prove the previous two statements.

....

Since the points $f^k(h)$ converge to p as $k \to \infty$, there are an infinite number of homoclinic points in any neighborhood of p. *Thus even a small neighborhood of the fixed point is not correctly described by the linear approximation.* $W^s(p)$ and $W^u(p)$ fold around to return repeatedly and intersect in every neighborhood of p, no matter how small. This *homoclinic tangle* leads to chaos. A way to comprehend chaos is to determine the points (e.g. near p) that are not driven off far away under the repeated action of f. The points on this *invariant set* Λ is mapped into each other by f, and so provide a subsystem. In chaotic systems, Λ is uncountably infinite, a sort of fractal. Its Hausdorff dimension is one way to quantify the chaos.

A.3   Arnold's cat map

Consider the linear map

$$x \mapsto \begin{pmatrix} 2 & 1 \\ 1 & 1 \end{pmatrix} \begin{pmatrix} x^1 \\ x^2 \end{pmatrix}.$$

It is clear that $x = (0,0)$ is a fixed point. The eigenvalues are $\frac{3\pm\sqrt{5}}{2} \approx 2.61803, 0.381966$. The first one is unstable and the second stable. The stable manifold is the set of vectors proportional to the eigenvector $\left(\frac{1-\sqrt{5}}{2}, 1\right)$ of the smaller eigenvalue and the unstable manifold is spanned by the other eigenvector $\left(\frac{1+\sqrt{5}}{2}, 1\right)$. So far we have a system with no chaos: just a fixed point of hyperbolic type.

Now, given any real number x, there is an integer m such that $|x - m| \leq 0.5$. Call this $x - m$ the fundamental part of x, denoted by $[x]$. Clearly, it is unchanged under translations by integers and is in the fundamental region $[-0.5, 0.5]$. The endpoints of the fundamental region are to be thought of as glued together, so that it represents the circle. Arnold's cat map[2] is defined as

$$f(x) = \begin{pmatrix} [2x^1 + x^2] \\ [x^1 + x^2] \end{pmatrix}.$$

[2] Arnold illustrated chaos by the effect of this map on an image of a cat under repeated applications; hence the name. Try it, but on a picture of your least favorite professor.

This is a smooth map of the torus to itself, thought of as the square $[-0.5, 0.5] \times [-0.5, 0.5]$ with opposite edges glued together. Again, $(0,0)$ is a fixed point of saddle type. Stable and unstable manifolds are curves that wind around the torus:

$$W^u = \left\{ \left(\left[\frac{1 + \sqrt{5}}{2} t \right], [t] \right) \mid t \in \mathbb{R} \right\},$$

$$W^s = \left\{ \left(\left[\frac{1 - \sqrt{5}}{2} t \right], [t] \right) \mid t \in \mathbb{R} \right\}.$$

If we represent the torus as a square, they start as straight lines at the fixed points. If you extend them out to large values of $|t|$, they will hit the boundary. By the periodic boundary condition, they must be continued from the opposite side as parallel straight lines. See Figure A.1.

If we extend the manifolds out further still, W^s and W^u will intersect. There are homoclinic points, whenever

$$\left[\frac{1 + \sqrt{5}}{2} t \right] = \left[\frac{1 - \sqrt{5}}{2} t \right], \implies t = \frac{m}{\sqrt{5}}, \quad m \in \mathbb{Z}.$$

The first such point is at $t_1 = \frac{1}{\sqrt{5}} \approx 0.447214$. Its forward image lies on the stable manifold. If we extend the unstable manifold eventually it will intersect it, since it is also a homoclinic point.

In this example there are so many intersections of W^s and W^u that the homoclinic points form a dense subset of the torus: see Figure A.1.

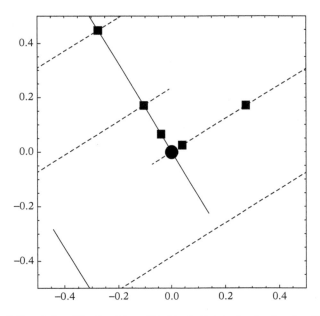

Figure A.1 *Manifolds of Arnold's Cat Map: The black disk is the fixed point. The solid lines form part of the stable manifold W^s; the dashed lines the unstable manifold W^u. Notice how they shows up at the opposite side after hitting a boundary. The square-shaped points are some of the homoclinic points, where the stable and unstable manifolds intersect*

A.4 The homoclinic tangle

It is useful to identify a neighborhood P of the fixed point which, when under some number of iterations (say m) will contain h, that is, $p \in P, h \in f^m(P)$.

You can visualize $f^m(P)$ as a "thickening" of the unstable manifold $W^u(p)$ extended just far enough to include h. Now, there is some number $n > 0$ such that $f^{-n}(P)$ contains h as well. This $f^{-n}(P)$ is a "thickening" of a part of the stable manifold $W^s(p)$. Then the iterate $\phi \equiv f^{m+n}$ will map $\phi : f^{-n}(P) \to f^m(P)$.

The situation is illustrated in Figure A.2. P is some region containing the fixed point (the black disk). When acted on $m = 2$ times by f, the region P is stretched along the unstable direction to get a region $X = f^m(P)$. P is chosen to be big enough that X contains h(the rectangle). With $n = 1$, the inverse cat map f^{-1} stretches P along the stable direction till it just contains h as well. This $f^{-1}(P)$ is the box labeled Q. Thus $\phi = f^{m+n}$ maps the region Q to X. It takes some fine tuning of the P to get this picture just right.

Now let us consider the dynamics of ϕ. Call $Q = f^{-n}(P)$ and $X = f^m(P)$. So $\phi : Q \to X$. Our aim is to understand the subset Λ of points in Q that remain in Q after many iterations of ϕ or ϕ^{-1}. These are the points that are mapped into each other by f without being driven off too far away: Λ is the chaotic set of ϕ. (For a large number of iterations, there is not much of a difference between looking at iterations of ϕ and those of f.)

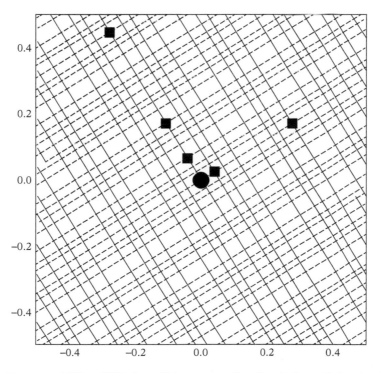

Figure A.2 *If we extend W^s and W^u they will intersect so often that the homoclinic points form a dense subset. A glimpse of a homoclinic tangle*

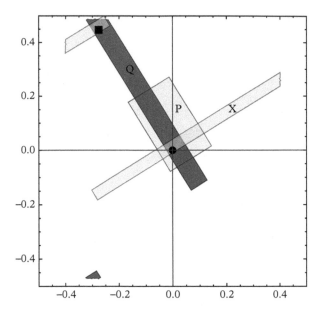

Figure A.3 *The horse shoe of the cat map: The black disk is the fixed point and the rectangle is a homoclinic point. The region Q is mapped to X by some iterate $\phi = f^{m+n}$*

Most of the points in Q are pushed out of it by ϕ. The intersection $Q \cap \phi(Q)$ consists of a smaller box around the fixed point and another box that contains the homoclinic point (Figure A.3), where the shaded regions overlap. These are the only points that remain in Q after the action of ϕ. If we hit with ϕ again, it will break up each of these two regions into two, so that $Q \cap \phi(Q) \cap \phi^2(Q)$ consists of four smaller pieces. We can also look in the other direction: $\phi^{-1} : X \to Q$ so that $\phi^{-1}(Q) \cap Q$ is also made of two disconnected regions, $\phi^{-1}(Q) \cap Q \cap \phi(Q)$ is the union of four such regions and so on. Each step removes more and more points from the intersection. The fraction removed is roughly constant, so the limiting set

$$\Lambda = \lim_{k \to \infty} \phi^{-k}(Q) \cap Q \cap \phi^k(Q)$$

is a fractal of dimension less than two: a sort of Cantor set. The exact shape of Q and X depend on the details of the dynamics f.

The important thing is that $X = \phi(Q)$ twists around and intersects with Q. With some imagination you can see that the picture can be simplified without losing the essence: the torus can be replaced by a square and the X can be replaced by a horse shoe shaped region. We just have have to design a smooth function ϕ that will map a square to such a horse shoe.

The fractal dimension of Λ only depends on the eigenvalues of $f'(p)$ at fixed point: there is a universality, an independence on the details of the dynamics f. We now turn to a deeper study of this fractal.

A.5 The Smale horse shoe

The simple harmonic oscillator is the basic model for a vibrating system; Kepler's problem is the basis for celestial mechanics. The horse shoe map of Smale plays a similar role for chaotic systems.[3] Every dynamical system with a homoclinic tangle contains a map equivalent to the horse shoe.

We want to understand the behavior of a diffeomorphism $f : M \to M$ near some hyperbolic fixed point p, after a large number of iterations.

Take a small disk P containing p and look at its image after some number of iterations. You can think of P as a fattened version of the fixed point: after all, physically there would always be some error in locating points and the best we can do is to be close to the intended point.

The image $f^k(P)$ will be elongated along the unstable direction and compressed along the stable direction; similarly $f^{-k}(P)$ will be stretched along the stable direction and shrunk along the unstable direction. If k is not too big, the picture will look like a cross centered at p. The simplest possibility is that as k grows, the images $f^k(P)$ and $f^{-k}(P)$ will stretch out further and never return to anywhere near p. In this case there is no chaos (at least near p): the system behaves predictably near p by either falling into p or getting kicked far away from p.

But suppose that for some value of k, $f^k(P)$ is shaped like a horse shoe ∩: it folds are returns to near p. Then $f^{-k}(P)$ will be shaped like a sideways horse shoe ⊃. Once the stable and unstable manifolds intersect, there are homoclinic points. These will accumulate at the fixed point: the manifolds are extremely convoluted there and intersect each other many times: the homoclinic tangle.[4]

Let $\phi = f^k$, h_1 the upper left homoclinic point and h_0 the lower right one. We focus on a region Q containing the tangle: the square with vertices p, h_0 and $\phi(h_1) = h_0$. Choose coordinates so that p is at the origin and the further iterates $\phi(h_0), \phi^2(h_0) \cdots$ tends to p along the lower horizontal strip.

A.6   Binary code

To summarize, we now have a diffeomorphism with a fixed point $p \in Q$, and homoclinic points $h_1 \in W^u(p) \cap Q$ and $h_0 \in W^s(p) \cap Q$ such that $\phi(h_0) = h_1$.

Our aim is to understand the invariant set $\Lambda = \bigcap_{k=1}^{\infty} \phi^{-k}(Q) \cap Q \cap \phi^k(Q)$ of the dynamics generated by ϕ. By its definition, ϕ shuffles the points of Λ among each other. We want a way to label the points of Λ which makes the action of ϕ transparent. This is analogous to the *normal coordinates* of a harmonic oscillator. A useful first step is a "binary code" for orbits invented by (Smale, 1967), perhaps inspired by Cantor's work on fractals.

We can split $Q = B_1 \sqcup Y \sqcup B_0$ into three disjoint but connected regions

- B_1 containing h_1
- B_0 containing p and h_0
- the remaining region Y such that $\phi(Y)$ is outside Q.

Moreover we can arrange that all the forward as well as backward iterates of h_1 are in B_0.

---

[3] Legend has it that Smale was hanging out at a beach in Brazil, while supported by an NSF fellowship, when he discovered this. Some Congressmen objected.

[4] Of course, more complicated things can happen: $f^k(P)$ might re-enter Q several times. The basic idea of symbolic dynamics described below remains unchanged, except with one symbol for each connected region in $\phi^{-1}(Q) \cap Q \cap \phi(Q)$.

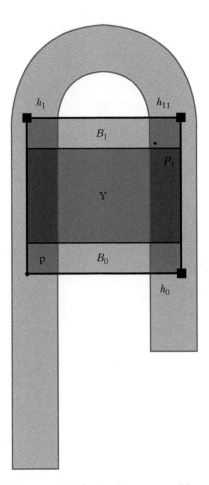

Figure A.4 *The horse Shoe map with fixed points p, p_1 and homoclinic points h_0, h_1, h_{11}*

We see now that the part of Q that is mapped to Q consists of two disconnected pieces B_0, B_1. We assign them numbers $0, 1$ respectively (Figure A.4).

In the next iteration, we have points of four types: $1.1 = \{x | x \in B_1, \phi(x) \in B_1\}$, then $1.0 = \{x | x \in B_1, \phi(x) \in B_0\}$ and similarly $0.1 = \{x | x \in B_0, \phi(x) \in B_1\}$ and finally $0.0 = \{x | x \in B_0, \phi(x) \in B_1\}$. The effect of acting with ϕ again will to split these regions further into two so that $Q \cap \phi(Q) \cap \phi^2(Q)$ has eight regions with labels such as $1.10 = \{x | x \in B_1, \phi(x) \in B_1, \phi^2(x) = B_0\}$ etc.

In the same way, $10. = \{x | \phi^{-1}(x) \in B_1, x \in B_0,\}$ and so on. Proceeding this way, we get $\Lambda_k \equiv \bigcap_{l=1}^{k} \phi^{-l}(Q) \cap Q \cap \phi^l(Q)$ is the union of 2^{2k} regions labeled by strings of 0 and 1 with k entries to the left of the point and k entries to the right. As k grows each connected piece of Λ_k shrinks to a point, which is labeled by a sequence of 0 and 1 extending to infinity both to the left and to the right.

This limit $\Lambda = \bigcap_{l=1}^{\infty} \phi^{-l}(Q) \cap Q \cap \phi^{l}(Q)$ is the invariant set of ϕ. It is in 1:1 correspondence with the space of bi-infinite sequences, of zeroes and ones: $\{0, 1\}^{\mathbb{Z}}$. The action by ϕ is equivalent to shifting the sequence to the left; the action by ϕ^{-1} shifts to the right. Just as the binary representation of numbers allows multiplication by 2 easily (shift the binary point), our encoding of points on Λ allows us to study the dynamics of ϕ easily.

In particular, we now see that there are at least as many points in Λ as there are points in the unit square. The numbers to the right of the period will represent, in binary, some point in the interval $[0, 1]$. The binary sequence to the left, read backward, of the "." gives another point in $[0, 1]$. Together, they determine the Cartesian co-ordinates of some point in the unit square. Reversing this construction, we can produce some point in Λ from the coordinates of any point in the unit square. Thus, *the invariant set Λ is uncountably infinite.*

But it has zero area. Each Λ_k consists of 2^{2k} regions, each of whose area decreases exponentially. Thus, Λ is a *fractal* (We will quantify this later in terms of the Hausdorff dimension.)

Example A.2

Clearly the fixed point has binary code $p = \cdots 0.00 \cdots$, $h_1 = \cdots 001.00 \cdots$ and $h_0 = \cdots 00010.000 \cdots$. We now discover that there must be another fixed point $p_1 = \cdots 1.111 \cdots$. It is located near the homoclinic point on the upper right hand corner. See Figure A.4.

Periodic points correspond to repeating sequences such as $\cdots 011.011011011 \cdots$. Any sequence can be approximated by a long enough periodic sequence: just cut out a long enough piece of it and repeat it over and over.

Theorem 1 *Periodic points are dense in Λ.*

The reason for chaos is that most points in Λ correspond to sequences that hop between 0 and 1 with no pattern whatever. Just as most real numbers are irrational, with no regularity in their binary expansion. To go deeper we will need some tools from the theory of fractals. It is best to develop them in the simpler context of subsets of an interval (rather than a square).

A.7   Iterated function systems on the interval

A standard method of constructing fractals is using an *iterated function system* (IFS). They are widely used in computer graphics to paint such things are trees, mountains and clouds which have a fractal appearance. We illustrate this idea with subsets of the unit interval. See (Falconer, 2003) for a dedicated discussion.

Suppose $S_{0,1} : [0, 1] \to [0, 1]$ are a pair of affine maps of the interval $I = [0, 1]$:

$$S_0(x) = \mu_0 x, \quad S_1(x) = 1 - \mu_1 x, \quad \mu_0, \mu_1 > 0, \quad \mu_0 + \mu_1 < 1.$$

Clearly, S_0 shrinks the unit interval to $[0, \mu_0]$, keeping 0 fixed and taking 1 to μ_0. Also, S_1 maps the unit interval to $[1 - \mu_1, 1]$ but it reverses the orientation in the process: it is 0 that gets mapped to 1 and 1 goes to $1 - \mu_1$. (This is a bit different from the standard choice, but suits our situation better.)

Let $F_1 \subset [0, 1]$ be the union of the images of the unit interval under the two maps above

$$F_1 = S_0(I) \cup S_1(I)$$

As we saw, it is a disconnected set, with one piece on the left and another on the right. Let us label the left component $S_0(I)$ by .0 and the right $S_1(I)$ component by .1 (the reason for putting the point in front will become clear soon):

$$.0 \leftrightarrow S_0(I) = [0, \mu_0], \quad .1 \leftrightarrow S_1(I) = [1 - \mu_1, 1].$$

We can also think of F_1 as obtained from I by removing the open interval $]\mu_0, 1 - \mu_1[$ in the middle. Define

$$F_2 = S_0(F_1) \cup S_1(F_1).$$

(Equivalently, F_2 is obtained from F_1 by removing an open interval in the middle from each component of F_1.)

This F_2 now has four connected pieces, each of the previous pieces having been split into a left and a right component.

We can label the left and right components of .0 by .00 and .01 respectively; and similarly .10 and .11 for the left and right components of .1. Each component has length less than μ^2 where μ is the larger of μ_0 or μ_1:

$$\mu = \max\{\mu_0, \mu_1\}.$$

Explicitly,

$$.00 \leftrightarrow S_0 S_0(I) = [0, \mu_0^2], \quad .01 \leftrightarrow S_0 S_1(I) = [\mu_0(1 - \mu_1), \mu_0],$$
$$.10 \leftrightarrow S_1 S_1(I) = [1 - \mu_1, 1 - \mu_1 + \mu_1^2] \quad .11 \leftrightarrow S_1 S_0(I) = [1 - \mu_0 \mu_1, 1].$$

After r steps

$$F_r = S_0(F_{r-1}) \cup S_1(F_{r-1}).$$

F_r will have 2^r components, each of length less than μ^r (recall that μ is the larger of μ_0 or μ_1). Thus the total length of intervals in F_r is less than $2^r \mu^r$. Since $\mu_0 + \mu_1 < 1$, we have that $2\mu < 1$. So the total length of F_r decreases exponentially with r.

In the limit $r \to \infty$ we will get a set that is invariant under the process of subdivision

$$F = S_0(F) \cup S_1(F).$$

Its components are points which are totally disconnected from each other: in the limit $r \to \infty$ each connected component shrinks to a point. Not only that, the total length of F is zero as well.

And yet there are an uncountably infinite number of points in F: to each infinite binary sequence there is a point in F. And there are as many such sequences as there are points on $[0, 1]$ (just think of the binary expansion of a number). The infinite binary sequence associated to a point in F determines a real number in $[0, 1]$. It may be thought of as a coordinate system on F.

Example A.3

The most famous example is the "middle third" Cantor set with $\mu_0 = \mu_1 = \frac{1}{3}$.

Although its length is zero, we can find a new measure of size for sets such as F. Define the "Hausdorff measure" of an interval of length l to be l^d for some number d. The power d is much like a dimension: the volume of a d-dimensional cube of side l would have been l^d. But we allow for d to be a fractional number, $0 < d < 1$. This will magnify the measure of small intervals, making sets such as F measurable. We just have to choose the d for which the measure of the two child sets from the action of f_0 or f_1 add up to *exactly* the measure of the parent. Since the lengths shrink by a factor of μ_0 (for the left component) and μ_1 (for the right component) the measures change by μ_0^d and μ_1^d, we get the equation that determines d:

$$\mu_0^d + \mu_1^d = 1.$$

This d is the Hausdorff dimension of the fractal F.

Example A.4

For the "middle third" Cantor set $\mu_0 = \mu_1 = \frac{1}{3}$ the Hausdorff dimension is given by $2 \times 3^{-d} = 1 \implies d = \frac{\log 2}{\log 3} \approx 0.63093$.

A.8   Normal form of the horse shoe

We can now get a more detailed picture of the invariant set Λ of the horse shoe map in the immediate neighborhood of the fixed point. A basic principle of calculus is that near a fixed point, the linear approximation to a differentiable map should be a good one. But, as we saw, this is only true for a finite number of iterations of the map. If we want to know what happens over a large number of iterations (which is needed to understand Λ), we will have to take nonlinearities into account. A small deviation from the fixed point will move along the unstable manifold. If it eventually returns to the immediate vicinity of the fixed point, the linear approximation cannot describe the resulting tangle anymore.

If the map is topologically similar to the horse shoe (shaped like a $\cap$, with a single fold), *a pair of linear approximations* can describe it well. Recall that the binary code of the original fixed point (let us call it p_0 now) is $\cdots 00.00 \cdots$. Also, there is another fixed point p_1 near p, with binary code $\cdots 11.11 \cdots$. We can approximate the map ϕ by a pair of affine transformations

$$A_0(x) = \mathcal{J}_0(x - p_0) + p_0, \quad A_1(x) = \mathcal{J}_1(x - p_1) + p_1,$$

where $\mathcal{J}_0, \mathcal{J}_1$ are the Jacobian matrices at the fixed points

$$\mathcal{J}^i_{0j} = \left[\frac{\partial \phi^i}{\partial x^j}\right]_{x=p_0}, \quad \mathcal{J}^i_{1j} = \left[\frac{\partial \phi^i}{\partial x^j}\right]_{x=p_1}.$$

A_0 is a good approximation in some region B_0 containing p_0 and A_1 in some B_1 containing p_1. We can now choose a coordinate system in B_0 in which A_0 is diagonal and p_0 is at the origin:

$$A_0(x,y) = \begin{pmatrix} \mu_0 x \\ v_0^{-1} y \end{pmatrix}, \quad 0 < \mu_0, v_0 < 1.$$

Within B_0, under each action of ϕ the stable direction x shrinks by a factor of μ_0 and the unstable direction y expands by a factor of v_0^{-1}. Also, $\phi(B_0) \subset Q$. We can thus identify

$$B_0 = \{(x,y) \mid 0 \le x \le 1, 0 \le y \le v_0\}$$

and

$$Q = \{(x,y) \mid 0 \le x \le 1, 0 \le y \le 1\}.$$

If $v_0 < y$, we are taken outside the square Q: this is the arch of $\cap$ shaped region.

Two non-commuting linear maps cannot be diagonalized in the same cartesian coordinate system. But we can have a system with linear axes in B_0, B_1 which are patched together by curvilinear extension (the fold) in the region Y between them. We do not need to find this extension, as the points in Y will not be part of the invariant set Λ. It is enough that it exists and is smooth.

So we now extend the coordinate system such that $\mathcal{J}_1$ is diagonal as well, and $h_0 = (0,1)$ and $h_1 = (1,0)$. Then,

$$A_1(x,y) = \begin{pmatrix} 1 - \mu_1 x \\ v_1^{-1}(1-y) \end{pmatrix}, \quad 0 < \mu_1, v_1 < 1.$$

The reason for the negative sign is the fold: in order for the two systems A_1 must not only scale the coordinates, but also rotate by half of a full turn. The translational terms are determined by the condition that A_1 map the homoclinic point h_0 to h_1:

$$A_1(h_1) = h_0 \implies (0,1) \mapsto (1,0).$$

Again, in order that $\phi(B_1) \subset Q$, we identify

$$B_1 = \{(x,y) \mid 0 \le x \le 1, 1 - \nu_1 \le y \le 1\}.$$

To patch together the two Cartesian coordinate systems in B_0, B_1, we will need a region Y in between them which is mapped outside Q: that is, B_0 and B_1 should not overlap. This requires

$$\mu_0 + \mu_1 < 1, \quad \nu_0 + \nu_1 < 1.$$

The region Y in between is mapped by ϕ to the fold, which is outside Q. Further iterations will drive these points out to infinity. The above condition on the eigenvalues requires that the two fixed points are "hyperbolic enough."

To orient ourselves, we can check that the other fixed point is at

$$p_1 = \left(\frac{1}{1+\mu_1}, \frac{1}{1+\nu_1}\right).$$

So, to study the dynamics of ϕ near p_0 and p_1, we can replace it with a "skeleton" consisting of mainly of a pair of affine maps:

$$A(x,y) = \begin{cases} A_0(x,y) & (x,y) \in B_0 \\ A_1(x,y) & (x,y) \in B_1 \\ \infty & \text{otherwise} \end{cases}.$$

Each iteration of ϕ (or its skeleton A) drives more and more points out of Q: each time a point falls in Y it is expelled to ∞. Only the unstable direction (the y-axis) matters in identifying the points expelled by the forward iteration. When restricted to the y-coordinate, the maps reduce become just $y \mapsto \nu_0^{-1} y$ and $y \mapsto 1 - \nu_1 y$. After the first iteration the points that remain are B_0 and B_1, that is, the image of the unit interval under

$$T_0(y) = \nu_0 y, \quad T_1(y) = 1 - \nu_1 y. \tag{A.1}$$

This is exactly how we constructed the Cantor set. So the surviving points will have y-coordinates belonging to the Cantor set $C(\nu_0, \nu_1)$ of the IFS of eqn (A.1).

If we now consider the inverse map, we will see that points get expelled depending on their x-coordinate. The corresponding IFS is

$$S_0(x) = \mu_0 x, \quad S_1(x) = 1 - \mu_1 x.$$

The surviving points have x-coordinates belonging to the Cantor set $C(\mu_0, \mu_1)$. Thus the invariant set is the Cartesian product of two Cantor sets:

$$\Lambda = C(\mu_0, \mu_1) \times C(\nu_0, \nu_1).$$

Just as the plane is a Cartesian product of two real lines, Λ is the product of a pair of Cantor fractals. The Hausdorff dimension of Λ is $d = d_x + d_y$ where d_x and d_y are the dimensions of the Cantor sets in the x and y coordinates:

$$\mu_0^{d_x} + \mu_1^{d_x} = 1, \quad \nu_0^{d_y} + \nu_1^{d_y} = 1.$$

Thus, the eigenvalues of the derivative of ϕ at the two fixed points fully determine the invariant set Λ and the dynamics on it. The coordinate system above is analogous to the normal coordinates of the harmonic oscillator: the dynamics is made as simple as possible in this system.

Given any binary sequence s, there is a periodic orbit of A corresponding to it. One point on this orbit can be found as the fixed point of the corresponding product A_s of affine transformations A_0, A_1; the remaining points on the orbit are found by the action of A. The invariant set Λ consists of all these periodic orbits and their limit points. (That is, Λ is the closure of the set of periodic orbits.) Thus, we can find the invariant set explicitly in the normal coordinates.

Example A.5

Suppose

$$\mu_0 = \frac{1}{3}, \quad \mu_1 = \frac{1}{4}, \quad \nu_0 = \frac{1}{2}, \quad \nu_1 = \frac{1}{5}.$$

The normal form of the dynamics is

$$A(x,y) = \begin{cases} \left(\frac{x}{3}, 2y\right) & 0 \leq y \leq \frac{1}{2},\ 0 \leq x \leq 1 \\ \left(1 - \frac{x}{4}, 5(1-y)\right) & \frac{4}{5} \leq y \leq 1,\ 0 \leq x \leq 1 \\ \infty & \text{otherwise} \end{cases}$$

The binary sequence 01011 corresponds to the periodic orbit with binary code

$$\cdots 0101101011.0101101011 \cdots$$

and its cyclic permutations. The product of affine transformations corresponding to it is

$$A_{01011} = A_0 A_1 A_0 A_1 A_1, \quad A_{01011}(x,y) = \left(\frac{5}{16} - \frac{x}{576}, 410 - 500y\right),$$

which has a fixed point at

$$(x_{01011}, y_{01011}) = \left(\frac{180}{577}, \frac{410}{501}\right).$$

Acting with A above, we get the orbit of period 5

$$\cdots \left(\frac{180}{577}, \frac{410}{501}\right), \left(\frac{532}{577}, \frac{455}{501}\right), \left(\frac{444}{577}, \frac{230}{501}\right), \left(\frac{148}{577}, \frac{460}{501}\right), \left(\frac{540}{577}, \frac{205}{501}\right), \left(\frac{180}{577}, \frac{410}{501}\right),$$
$$\left(\frac{532}{577}, \frac{455}{501}\right) \cdots$$

plotted in Figure A.5.

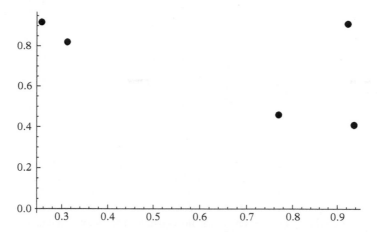

Figure A.5 *The periodic orbit of the sequence* 01011 *with parameters*
$$\mu_0 = \tfrac{1}{3}, \quad \mu_1 = \tfrac{1}{4}, \quad \nu_0 = \tfrac{1}{2}, \quad \nu_1 = \tfrac{1}{5}$$

Exercise A.3 Research project Write a program that will calculate the periodic orbit corresponding to a given binary sequence and parameters $\mu_0, \mu_1, \nu_0, \nu_1$. Use this to plot all the periodic points of period less than some given number k. This is an approximation to the invariant set Λ.

Appendix B
Chaotic Advection

The deepest questions of fluid mechanics have to do with understanding how chaos in the solution of Euler/Navier–Stokes equations leads to turbulence. But there is a simpler, more immediate way that chaos enters our subject. Each fluid flow—even static solutions of Euler/Navier–Stokes equations—defines a dynamical system: the advection of particles by that flow. Equivalently, the integral curves of the velocity field. Even simple flows can lead to chaotic dynamics. After all, even the iteration of a quadratic function is chaotic. See Appendix C. This fact, although obvious to people working in dynamical systems, was not elaborated upon in the context of fluid mechanics until (Aref, 1984). We will summarize his work in this short appendix.[1] It has led to direct experimental verification of chaos theory. And even to some useful fluidic devices.

B.1   Stirred fluid inside a cylinder (again)

We solved for the motion of an ideal incompressible fluid inside a cylindrical container of radius a, with a stirrer at ζ creating a vorticity Γ. The velocity field can be obtained by the method of images, as in Section 4.4.2. We saw that the integral curves are circles. But the rate at which the angle around the circle changes with time is not uniform. Now we return to this problem to determine the time dependence.

The problem is to solve the ordinary differential equation (ODE)

$$\frac{d\bar{\xi}}{dt} = \frac{\Gamma}{2\pi i}\left\{ \frac{1}{\xi - \zeta} - \frac{1}{\xi - \frac{a^2}{\zeta}}\right\}, \quad \xi(0) = z_0.$$

Since we know the shape of the orbit is a circle, it is natural to make the ansatz

$$\xi(t) = z_c + re^{i\theta(t)}$$

with a real angle θ. The center and radius of the circle are given by

$$z_c = \frac{\zeta - \lambda^2 \frac{a^2}{\zeta}}{1 - \lambda^2}, \quad r = \frac{\lambda}{1 - \lambda^2}\left|\frac{a^2}{\bar{\zeta}} - \zeta\right|.$$

---

[1] Note some minor changes of notation, for example, meanings of z and ζ are changed.

where λ is determined by the initial position of the advected particle:

$$\lambda = \left| \frac{z_0 - \zeta}{z_0 - \frac{a^2}{\zeta}} \right|.$$

The calculations are simplified if we choose the coordinate axes so that ζ is real. Then the ODE reduces to

$$\left[1 - \frac{2\lambda}{1 + \lambda^2} \cos \theta \right] \dot{\theta} = \frac{\Gamma}{2\pi r^2} \frac{1 - \lambda^2}{1 + \lambda^2}.$$

Integrating this

$$\theta - \frac{2\lambda}{1 + \lambda^2} \sin \theta = \frac{\Gamma}{2\pi r^2} \frac{1 - \lambda^2}{1 + \lambda^2} (t - t_0).$$

So the quantity

$$\Phi_\lambda(\theta) = \theta - \frac{2\lambda}{1 + \lambda^2} \sin \theta$$

varies at a constant rate. Since $0 < \lambda < 1$, this $\Phi_\lambda(\theta)$ is a monotonically increasing function. So it has an inverse $\Psi_\lambda(\phi)$ which can be determined numerically if necessary.

Thus, the integral curve is obtained by turning the initial point through an angle $\theta(t) = \Psi_\lambda\left(\frac{\Gamma}{2\pi r^2} \frac{1-\lambda^2}{1+\lambda^2} t \right)$ around an arc of a circle centered at z_c:

$$\xi(t) = z_c + [\xi(0) - z_c] e^{i\theta(t)}.$$

Since the particles move along in circles the fluid does not mix well; there is no chaos in this advection. Next we look at the case where the stirrer is allowed to change position.

B.2   Moving stirrers

In solving the equations for incompressibility and irrotationality, we did not use the information that the stirrer is at a fixed location. If its position and strength is given as a function of time we can get

$$v(z, t) = \frac{\Gamma(t)}{2\pi i} \left\{ \frac{1}{z - \zeta(t)} - \frac{1}{z - \frac{a^2}{\zeta(t)}} \right\}.$$

If the functions $\Gamma(t)$ and $\zeta(t)$ are periodic (with some period T) we have a machine with a rod that rotates and moves around to end up at the same point. By letting it run over many periods we might be able to mix the fluid completely. The engineering challenge is to design such a stirring protocol that is simple to implement and has rapid mixing or chatoic advection. There are many applications where fluid in a small confined region needs to be moved so that its ingredients mix quickly.

B.3   Aref's protocol

Aref proposed a simple protocol that has chaotic advection in some range of parameters. The vortex strength Γ is held constant and the position switches back and forth between two locations.

$$\zeta(t) = b \quad \text{for } nT \le t \le \left[n + \frac{1}{2}\right]T \tag{B.1}$$

$$= -b \quad \text{for } \left[n + \frac{1}{2}\right]T \le t \le [n+1]\,T.$$

In practice, we would insert two stirrers at b and $-b$; only the first stirrer is allowed to turn for a time $\frac{T}{2}$ and only the second for the next $\frac{T}{2}$ units of time. The dimensionless ratios $\mu = \frac{\Gamma}{2\pi a^2}T$ and $\beta = \frac{b}{a}$ determine the nature of the flow.

Let $f(z)$ be the point to which an initial particle in the particle is advected at the completion of one period; we will call it the Aref map. For half the time, the particle evolves around one circle; then suddenly it switches to a circle with a different center and radius determined by the new position of the stirrer. The combined effect is $f(z)$.

....

Exercise B.1 Write a program (e.g., in Mathematica) that numerically computes the Aref map $z \mapsto f(z)$ given μ, β. Plot orbits for randomly chosen initial conditions for 10^4 steps to see how the orbits fills most of the circle.

....

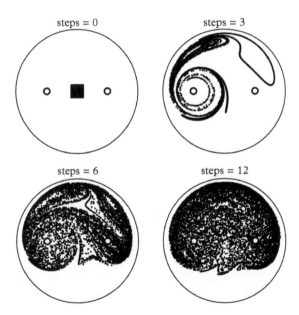

Figure B.1 *Chaotic Advection:* 10^4 *particles are initially spread uniformly on a square at the center of side 0.2. Their positions after various number of steps of time evolution are plotted. The stirrer is located at the right dot for half the time and at the left dot for the other half of each step*

The real point of Aref's work is that chaotic advection sets in very quickly, after just a few steps (not after thousands). So it is more instructive to look at how a small region near the origin evolves after a few steps. We plot in Figure B.1 how the points initially within a square of side 0.2 centered at the origin evolve in a few steps (for the choice of parameters $\mu = 2.0, \beta = 0.5$).

....

Exercise B.2 It is a theorem of topology that every continuous map of the disk to itself (which preserves the boundary) must have a fixed point. Find (using the numerical secant method) a fixed point with $\mu = 2.0, \beta = 0.5$. Calculate numerically the Jacobi matrix at the fixed point. Verify that it has determinant one. (Why?) Find the stable and unstable axes at the fixed point.

....

....

Exercise B.3 Research project Is there a horse shoe inside the Aref map? The stable and unstable manifolds ought to intersect if they are extended far enough. Study the resulting homoclinic tangle using methods of the last chapter.

....

Appendix C
Renormalization

Renormalization is a technique for deducing the properties of large systems from those of small constituents. It arose in studies of point particles in quantum electrodynamics (Dirac, 1938; Schwinger, 2003) and was further developed in elementary particle physics ('t Hooft, 1994).

It was not realized at first that the same ideas are useful in statistical physics, where also the problem is to deduce the large scale effects from interactions of very small things like molecules. A phase transition is a spectacular phenomenon: a discontinuous jump in energy, like the latent heat of water as it boils to steam. This is called a first order transition. As we increase the pressure, the latent heat decreases.

At a critical pressure and temperature, the latent heat vanishes so that the energy is continuous. But its derivative (specific heat) is infinite: this is called a second order phase transition. It turns out that the mathematical theory of this divergence is exactly the same as of the divergences which previously occurred in particle physics (Kleinert and Schulte-Frohlinde, 2001). So, these are not different subjects to be pursued by warring factions of physicists: instead, they are two manifestations of the same phenomenon.

It was already known that the equation of state of all materials are the same near a second phase transition (universality). For example, magnets and liquids have specific heats that diverges as $(T - T_c)^{-\alpha}$ with the same critical exponent; predicted theoretically to be $\alpha \approx 0.1096$. (Experimentally, $\alpha = 0.1105 + 0.025 \pm 0.027$.) Why would such completely different physical systems behave the same way near a critical point? This was a big puzzle of twentieth century physics. Wilson's explanation of this universality is now the most important application of renormalization theory: (Kleinert and Schulte-Frohlinde, 2001).

The transition from laminar to turbulent flow has many analogies to a phase transition. As early as 1941, Kolmogorov discovered a scaling law which remains the foundational work of modern theory of turbulence (Davidson, 2004). Many of us therefore suspect that renormalization is the key to understanding turbulence. Alas, this research lies outside the scope of this book.

In this appendix we will introduce renormalization in contexts more elementary than fluid mechanics, where the ideas are easier to work out explicitly. The purpose is not to chart this vast and deep ocean of ideas. But to visit a few accessible islands and encourage some of the readers to undertake the necessary voyages of exploration. Every physicist needs to know at least this much about it.

C.1 The Ising model

Our aim here is to build the simplest model of a large number of molecules with magnetic moments. The collective behavior of these magnets must lead to a low temperature phase in which most moments point in the same direction (the ordered phase). At high temperatures, there should be a disordered phase in which the magnetic moments average to zero. A molecule has a magnetic moment that is proportional to its angular momentum (spin). The proportionality constant (the gyromagnetic ratio) depends on the structure of the molecule and is not important for us.

In the simplest model, the Ising model, this magnetic moment (or spin) can point in one of two directions: we have a variable $\sigma = \pm 1$ at each molecule that describes this orientation. There are more intricate models where the spin can lie on a circle (the XY model) or on a sphere (the Heisenberg model) but we stick with the simplest case. The molecules are arranged on a cubic lattice in d dimensions, of L sites in each direction. (Other graphs can be chosen as well.)

Two neighboring spins will interact with an energy $-\tilde{\jmath}\sigma\sigma'$. If the constant $\tilde{\jmath} > 0$, the spins will have a tendency to align producing a ferromagnet at low temperatures.[1] This interaction is due to an intricate quantum mechanical phenomenon involving tunneling (the exchange interaction) and decays exponentially with distance (much like chemical bonds). We can ignore the interaction except for molecules that are very close together. So we only include interactions among nearest neighbors.

Thus the magnetic energy of the Ising model is

$$H(\sigma) = -\tilde{\jmath}\sum_{x-y}\sigma_x\sigma_y,$$

where x–y denotes two positions on the lattice that are connected by a nearest neighbor bond. It is useful to include the possibility of an externally imposed magnetic field as well, so that

$$H(\sigma) = -\tilde{\jmath}\sum_{x-y}\sigma_x\sigma_y - \tilde{B}\sum_{x}\sigma_x.$$

The quantity $\tilde{B}$ is the magnetic field times the gyromagnetic ratio of the molecule. All the large scale (thermodynamic) behavior of the system can be deduced from the *partition function*

$$Z_L\left(\frac{\tilde{\jmath}}{k_BT},\frac{\tilde{B}}{k_BT}\right) = \sum_{\sigma=\pm 1} e^{\frac{\tilde{\jmath}}{k_BT}\sum_{x-y}\sigma_x\sigma_y + \frac{\tilde{B}}{k_BT}\sum_x\sigma_x},$$

where k_B is Boltzmann's constant and T is temperature. Since only the ratios appear we can use instead the variables

$$\jmath = \frac{\tilde{\jmath}}{k_BT}, \quad B = \frac{\tilde{B}}{k_BT}.$$

Recall that the thermodynamic free energy is the limit of a large number of molecules:

$$W(\jmath,B) = \lim_{L\to\infty} -\frac{\log Z_L(\jmath,B)}{L^d}.$$

From this every other thermodynamic quantity (specific heat, magnetization, etc.) can be calculated by standard formulas in thermodynamics (Landau and Lifshitz, 1980).

No one has been able to get an analytic formula for the free energy for the cubic lattice in three dimensions. It is commonly accepted that this is impossible in terms of the usual functions (elliptic, Painlevé etc.) known to mathematical physicists. A major advance was made by Onsager in the 1940s when he evaluated the sum for a square lattice in terms of elliptic functions. This

---

[1] If $\tilde{\jmath} < 0$ the spins will try to be opposite: an antiferromagnet.

showed for the first time that summing over molecular degrees of freedom can lead to singularities in the free energy: until then it was only a conjecture that phase transitions could be explained this way.

Onsager was building on a technique developed by Ising for the one-dimensional lattice, called the transfer matrix method. Although the one dimensional model is too simple (it does not have a phase transition) it is a good place to start.

C.2   Transfer matrix

Imagine a long chain of L spins arranged at regular intervals along a line. Each spin interacts with its two nearest neighbors. We will eventually take the limit $L \to \infty$. The hamiltonian can be written as

$$H = -\tilde{\mathcal{J}} \sum_{x=1}^{L-1} \sigma_x \sigma_{x+1} - \frac{\tilde{B}}{2} \sum_{x=1}^{L-1} (\sigma_x + \sigma_{x+1}).$$

It is convenient to split the term involving a single spin as an average of nearest neighbors.[2] The partition function is then

$$Z_L(\mathcal{J}, B) = \sum_{\sigma = \pm 1} e^{\mathcal{J}\sigma_1\sigma_2 + \frac{B}{2}(\sigma_1 + \sigma_2)} e^{\mathcal{J}\sigma_2\sigma_3 + \frac{B}{2}(\sigma_2 + \sigma_3)} e^{\mathcal{J}\sigma_3\sigma_4 + \frac{B}{2}(\sigma_3 + \sigma_4)} \cdots e^{\mathcal{J}\sigma_{L-1}\sigma_L + \frac{B}{2}(\sigma_{L-1} + \sigma_L)}$$

Define the 2×2 matrix labeled by $\sigma, \sigma' = \pm 1$:

$$P_{\sigma\sigma'} = e^{\mathcal{J}\sigma\sigma' + \frac{B}{2}(\sigma + \sigma')},$$

$$P = \begin{pmatrix} e^{\mathcal{J}+B} & e^{-\mathcal{J}} \\ e^{-\mathcal{J}} & e^{\mathcal{J}-B} \end{pmatrix}.$$

This is called the transfer matrix: it transfers us by one step along the lattice. Then

$$Z_L(\mathcal{J}, B) = \sum_{\sigma_1, \sigma_2 \cdots = \pm 1} P_{\sigma_1\sigma_2} P_{\sigma_2\sigma_3} P_{\sigma_3\sigma_4} \cdots P_{\sigma_{L-1}\sigma_L}.$$

The simplifying feature of the one-dimensional model is that the σ_2 only appears in the first two factors, σ_3 only in the second and third and so on. The sum over $\sigma_2, \sigma_3 \cdots \sigma_{L-1}$ can be thought of as matrix multiplication:

$$Z_L(\mathcal{J}, B) = \sum_{\sigma_1, \sigma_L} P_{\sigma_1\sigma_L}^{L-1}.$$

---

[2] The boundary spins are counted with half the strength; they don't matter much anyway in the limit $L \to \infty$.

The power of a symmetric matrix P^{L-1} can be calculated conveniently in terms of its eigenvalues (which are real) and eigenvectors (which can be chosen to be orthonormal):

$$P\psi_1 = \lambda_1\psi_1, \quad P\psi_2 = \lambda_2\psi_2,$$
$$\psi_1^T\psi_1 = 1 = \psi_2^T\psi_2, \quad \psi_1^T\psi_2 = 0,$$
$$P^L = \lambda_1^L\psi_1\psi_1^T + \lambda_2^L\psi_2\psi_2^T.$$

Suppose $|\lambda_1| > |\lambda_2|$,

$$P^L = \lambda_1^L\left[\psi_1\psi_1^T + \left(\frac{\lambda_2}{\lambda_1}\right)^L\psi_2\psi_2^T\right] \to \lambda_1^L\psi_1\psi_1^T$$

as $L \to \infty$. Thus the free energy is simply the log of the largest eigenvalue of the transfer matrix:

$$W = -\lim_{L\to\infty}\left[\log\lambda_1 + \frac{1}{L}\log\psi_1\psi_1^T\right] = -\log\lambda_1.$$

A little algebra gives

$$\lambda_1 = \frac{1 + v^2 + \sqrt{1 + (-2 + 4u^4)\,v^2 + v^4}}{2uv}$$

with

$$u = e^{-\mathcal{J}}, \quad v = e^{-B}.$$

The only singularities (poles or branch cuts) in the physical region $u, v \geq 0$ are at $u = 0, 1$ or $v = 0, 1$; these correspond to zero or infinite temperature. The one-dimensional Ising model does not have a phase transition at any intermediate values. Still it is a starting point for more intricate models. Now let us solve this method a different way.

C.3 Renormalization dynamics

Instead of summing over all the spins, in $Z_L(\mathcal{J}, B) = \sum_{\sigma_1,\sigma_2\cdots=\pm 1} P_{\sigma_1\sigma_2}P_{\sigma_2\sigma_3}P_{\sigma_3\sigma_4}\cdots P_{\sigma_{L-1}\sigma_L}$, suppose we sum over just half the spins (at odd numbered sites in the interior, 3, 5, 7, ...). We are then left a lattice with half as many sites, and the transfer matrix is the square of the previous one. By iterating this, we can eventually reduce the problem to a lattice of small size. (No harm in assuming that $L = 2^\Lambda$, a power of two.) The tough part then is to determine the effect of squaring the transfer matrix repeatedly.

It will be convenient to parametrize an arbitrary symmetric matrix (which has three independent elements) as

$$P = C\begin{pmatrix} \frac{1}{uv} & u \\ u & \frac{v}{u} \end{pmatrix}.$$

Thus, C is the fourth root of the product of all the elements; u is the off-diagonal element divided by this C; and v is the square root of the ratio of the diagonal elements. Physically, the parameter C

is the common energy of all the four spin configurations of the pair of sites. u describes a spin–spin interaction and v the interaction with an external field.

Then

$$P^2 = C^2 \begin{pmatrix} \frac{1}{u^2 v^2} + u^2 & v + \frac{1}{v} \\ v + \frac{1}{v} & u^2 + \frac{v^2}{u^2} \end{pmatrix} \equiv C_1 \begin{pmatrix} \frac{1}{u_1 v_1} & u_1 \\ u_1 & \frac{v_1}{u_1} \end{pmatrix}.$$

The effect of squaring the transfer matrix is a map of this three dimensional space to itself ("renormalization dynamics"):

$$(u, v, C) \mapsto (u_1, v_1, C_1),$$

$$u_1 = \frac{\left(v + \frac{1}{v}\right)^{\frac{1}{2}}}{\left(u^4 + u^{-4} + v^2 + v^{-2}\right)^{\frac{1}{4}}}.$$

$$v_1 = \frac{\left(u^4 + v^2\right)^{\frac{1}{2}}}{\left(u^4 + v^{-2}\right)^{\frac{1}{2}}},$$

$$C_1 = C^2 \left(v + v^{-1}\right)^{\frac{1}{2}} \left(u^4 + u^{-4} + v^2 + v^{-2}\right)^{\frac{1}{4}}.$$

Iterating this map amounts to summing over more and more spins. In the limit of a large number of iterations if we approach some fixed point, we can predict the result of an infinite sum over spins. Let us focus on u, v as C has no effect on their evolution. There is a fixed point at $(u, v) = (0, 1)$. Also, if $u = 1$, any value of v is a fixed point, that is, there is a line of fixed points $(1, v)$. The fixed point at $(0, 1)$ is unstable: if we start close to it, we are driven away from it at each iteration. No matter where we start, we will eventually end up somewhere on the line $(1, v)$: these are stable fixed points.

....

Exercise C.1 Do some numerical calculations of the orbit to verify these statements.

....

Any initial choice (except for $u = 0, v = 1$) tends after enough iterations to a system with a large external field and small interactions between spins. Mean field theory is this ultimate destination.

We can understand the orbits more systematically (not just numerical experiments) by using ideas from the theory of dynamical systems.

If there is a function $\Phi(u, v)$ that increases monotonically along each orbit ("Lyapunov function"), we can rule out chaotic or peridoic behavior for the orbits. A little thought will show that the loglog of the ratio of the largest to the smallest eigenvalue is such a function:

$$\Phi(u, v) = \log \log \frac{\lambda_1}{\lambda_2},$$

$$\Phi(u, v) = \log \log \left(\frac{1 + v^2 + \sqrt{1 + (-2 + 4 u^4) v^2 + v^4}}{1 + v^2 - \sqrt{1 + (-2 + 4 u^4) v^2 + v^4}} \right).$$

Exercise C.2 Show that under each iteration Φ increases by $\log 2$:

$$\Phi(u_1, v_1) = \Phi(u, v) + \log 2.$$

Hint: What happens to the eigenvalues as $P \mapsto P^2$?

We can visualize orbits better if we can find a function that is unchanged under iteration ("conserved quantity"): the orbits will lie along its contours. See Figure C.1. Since P and P^2 have the same eigenvectors, the ratio of the two components of an eigenvector of P is such a function (Taking the ratio gets rid of the normalization constant.):

$$\xi(u, v) = \frac{-1 + v^2 + \sqrt{[1 - 2v^2 + 4u^4v^2 + v^4]}}{2u^2v}.$$

The curves on which $\xi(u, v)$ is constant connect $(u = 0, v = 1)$ to some point on the line $u = 1$. Near the point $(u = 0, v = 1)$

$$\xi(u, v) \approx \frac{v - 1}{u^2} + 1,$$

while $\xi(1, v) = v$.

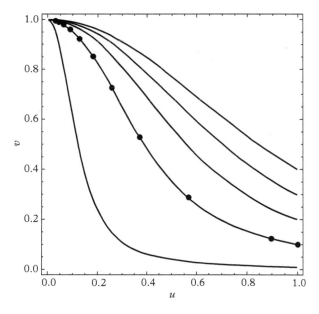

Figure C.1 *The dots are an orbit of the renormalization dynamics of the one-dimensional Ising model. The orbit starts at a region of small magnetic field and large interaction and tends towards small interactions and large magnetic field. The curves have constant $\xi(u, v)$*

A little thought will show that the initial points on a parabola $u_0^2 = \frac{1-v_0}{1-v_\infty}$ near the unstable fixed point tend to the final point $(1, v_\infty)$. In terms of the parameters we started with $B_0 = \left[1 - e^{-B_\infty}\right] e^{-2\mathcal{J}_0}$. Thus large spin–spin interactions $\mathcal{J}_0$ and exponentially small external fields B_0 evolve to vanishing spin–spin interactions with an external magnetic field.

We can think of B as the sum of a mean magnetic field created by the spins themselves plus any externally imposed ("bare") field. At each step of averaging, this effective magnetic field is enhanced, so that an exponentially small "bare" field B_0 grows to a finite value B_∞ over large scales. Simultaneously, the interaction between spins decreases to zero. Thus, spins with large mutual interactions and tiny bare fields evolve under averaging to free spins aligned along a mean field.

The Ising chain does not have spontaneous magnetization (non-zero average spin for zero external field). But it does provide a mechanism for enhancing any small stray field that may be present: such a field causes the alignment of more and more spins as they are averaged out to produce a large mean magnetic field in the end. So we are on the right track to finding a realistic model for magnets.

C.4 Cayley tree

Very few realistic physical systems can be solved exactly. We often solve some mathematical idealizations to get some insight into more realistic systems. One of these is the Ising model on a tree graph (Baxter, 2008). We start at some site labeled 0, and connect q sites to it. Then we add $q - 1$ sites to each of these so that they have q neighbors each (including the original root site). Then we repeat the process n times. The sites at the last step each have only one neighbor: they are at the boundary. This is called the Cayley tree of n generations (see Figure C.2). The limit $n \to \infty$ is an infinite tree graph where each site is connected to q others. (This is closely related to the Bethe lattice.) In the limit, any site can viewed as the root from which the whole graph is built as above.

If $q = 2$ this is simply the chain we described in the last section. When $q > 2$ the number of points at the boundary, $q(q - 1)^n$, increases exponentially with n. Moreover, the number of boundary points divided by the total number of points ("surface area divided by volume") tends to a constant as $n \to \infty$. These features make such a graph quite different from lattices in Euclidean space. Instead, a Cayley tree approximates hyperbolic geometry (a space of constant negative curvature). Still, these models give another simple mathematical model for renormalization.

Let us consider the Ising model on a Cayley tree of n generations. At each site we have a spin variable taking two possible values $\sigma_x = \pm 1$. The partition function is

$$Z_n = \sum_{\sigma=\pm 1} e^{\mathcal{J} \sum_{x-y} \sigma_x \sigma_y + B \sum_x \sigma}.$$

We can write this as

$$Z_n = \sum_{\sigma_0=\pm 1} e^{B\sigma_0} \left[g_n(\sigma_0)\right]^q,$$

where $g_n(\sigma_0)$ is the result of summing over all the spins on a tree graph with the spin at the root vertex held fixed at σ_0. Similarly, the average of σ_0 ("magnetization") is given by

$$M = \frac{\sum_{\sigma_0=\pm 1} \sigma_0 e^{B\sigma_0} \, [g_n(\sigma_0)]^q}{\sum_{\sigma_0=\pm 1} e^{B\sigma_0} \, [g_n(\sigma_0)]^q}.$$

If we remove the root site, the rooted Cayley tree of n generations breaks up into $q - 1$ rooted Cayley trees of $n - 1$ generations. We can use this to get a recursion relation:

$$g_n(\sigma_0) = \sum_{\sigma_1=\pm 1} e^{J\sigma_0\sigma_1 + B\sigma_1} \, [g_{n-1}(\sigma_1)]^{q-1}.$$

To solve these recursion relations, it is useful to define a new variable

$$x_n = \frac{g_n(-1)}{g_n(+1)}$$

as well as $u = e^{-J}, v = e^{-B}$ as before. Then the recursion becomes

$$x_n = f(x_{n-1})$$

with

$$f(x) = \frac{\frac{u}{v} + \frac{v}{u} x^{q-1}}{\frac{1}{uv} + uvx^{q-1}}.$$

The limit $n \to \infty$ will be given by the fixed point of the map $f(x)$. For each u, v we will look for a solution to the equation $x = f(x)$:

$$x = \frac{u^2 + v^2 x^{q-1}}{1 + u^2 v^2 x^{q-1}}.$$

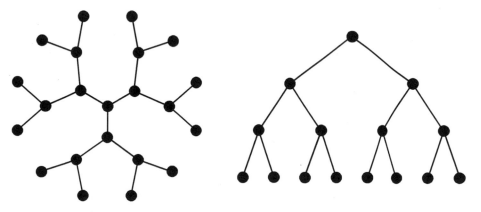

Figure C.2 *The diagram on the left is a Cayley tree. On the right is rooted Cayley tree; removing the top vertex (the root) breaks it into two smaller rooted Cayley trees*

This will then determine the magnetization which is, in the new variables,

$$M = \frac{v^{-2} - x^q}{v^{-2} + x^q}.$$

Of greatest interest is to see whether there is a magnetization even when the external field $B = 0$.

C.5   Spontaneous magnetization of the Ising model on the Cayley tree

To determine the spontaneous magnetization of the Ising model on the Bethe lattice, we put $v = 1$ (i.e., $B = 0$) and find the fixed point of the above map:

$$x = \frac{u^2 + x^{q-1}}{1 + u^2 x^{q-1}}.$$

Then the magnetization is given parametrically as a function of u through the two equations:

$$M = \frac{1 - x^q}{1 + x^q},$$

$$u = \sqrt{x \frac{1 - x^{q-2}}{1 - x^q}}.$$

By plotting it we see that there is indeed spontaneous magnetization for small enough temperatures. (Recall that $\mathcal{J} = \frac{1}{T}$ in some units, so that $u = e^{-\frac{1}{T}}$. So $T < T_*$ means that $u < u_*$ for some transition value u_*). Taking the limit $x \to 1$ we see that the transition happens at $u_* = \sqrt{\frac{q-2}{q}}$. As expected, this is uninteresting ($T_* = 0$) for $q = 2$, the Ising model on the chain. Expanding near the transition,

$$u \approx u_* \left[1 - \frac{1}{12}(q-1)(1-x)^2 \right] + O\left([1-x]^3 \right),$$

$$M \approx \frac{1}{2} q(1-x) + O\left([1-x]^2 \right)$$

so that

$$1 - x \propto \sqrt{u_* - u},$$

$$M \propto (u_* - u)^{\frac{1}{2}}.$$

Thus the spontaneous magnetization (Figure C.3) vanishes near the transition point like $M \sim (u_* - u)^\beta$ with an exponent $\beta = \frac{1}{2}$ independent of q (as long as $q > 2$). This does not agree with the two- or three-dimensional Ising models in Euclidean space; instead it agrees with the mean field approximation.

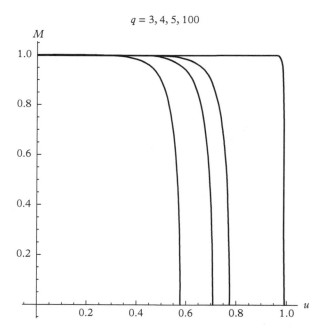

$$q = 3, 4, 5, 100$$

Figure C.3 *The spontaneous magnetization of the Ising model on a Cayley tree for various coordination numbers q, as a function of $u = e^{-\mathcal{J}}$. As q increases, the transition occurs for smaller values of the interaction $\mathcal{J}$*

Exercise C.3 Determine the spontaneous magnetization of the Ising model on a Bethe lattice with $q = 4$. Find the transition point. Compare with the critical point of the Ising model on a square lattice at $\mathcal{J} = \frac{\log[1+\sqrt{2}]}{2}$.

C.5.1 Specific heat

To get the free energy we need to know $g_n(\sigma)$ itself, not just the ratio of its two values. The extra information can be encoded into a variable y_n:

$$g_n(-1) = e^{y_n}\sqrt{x_n}, \quad g_n(+1) = \frac{e^{y_n}}{\sqrt{x_n}}.$$

The recursion becomes

$$e^{y_n}\sqrt{x_n} = e^{(q-1)y_{n-1}}\left[\frac{v}{u}x_{n-1}^{\frac{q-1}{2}} + \frac{u}{v}x_{n-1}^{-\frac{q-1}{2}}\right].$$

Since we already know that x_n approaches a limiting value x as $n \to \infty$ we can put that in to simplify the relation for large n:

$$y_n = (q-1)y_{n-1} + z, \quad z = \log\left[\frac{v}{u}x^{\frac{q}{2}-1} + \frac{u}{v}x^{-\frac{q}{2}}\right].$$

The solution is

$$y_n = \frac{(q-1)^{n-1}-1}{(q-1)-1}z + (q-1)^{n-1}y_1.$$

We can evaluate the case of a tree with a single generation "by hand" to get

$$y_1 = \frac{q-1}{2}\log\left[u^2 + \frac{1}{u^2} + v^2 + \frac{1}{v^2}\right].$$

In the limit $n \to \infty$ the number of vertices grows like $(q-1)^n$. The free energy per site is

$$F = -\lim_{n\to\infty}\frac{1}{(q-1)^n}\log Z_n.$$

In the large n limit, the differences between $\log Z_n$, $\log g_n(-1)$, $\log g_n(+1)$ are too small to matter. So we get

$$F = -\lim_{n\to\infty}\frac{1}{(q-1)^n}\log g_n(-1)$$

$$= -\frac{1}{(q-1)^2}z - y_1.$$

To find the specific heat we must use parametric differentiation. Remembering $u = e^{-\frac{1}{T}}$,

$$C = \frac{dF}{dT} = \frac{\frac{dF}{dx}}{\frac{du}{dx}}\frac{du}{dT} = u(\log u)^2\frac{\frac{dF}{dx}}{\frac{du}{dx}}.$$

At the transition point $x \to 1$, both $\frac{dF}{dx}$ and $\frac{du}{dx}$ vanish like $(1-x)$ so that the ratio is finite. Thus the specific heat remains finite at the transition. Thus the Ising–Cayley model does not have a second order phase transition. It behaves like mean field theory.

....

Exercise C.4 Study the Potts model on a Cayley tree with coordination number q. When is there a spontaneous magnetization? Where is the phase transition? What is the behavior near the transition (critical exponents)?

....

The Potts model is a generalization of the Ising model. The "spin" takes n possible values. The hamiltonian is $H = -\mathcal{J}\sum_{x-y}\delta(\sigma_x,\sigma_y)$ where $\delta(\sigma,\sigma') = 1$ if $\sigma = \sigma'$ and $\delta(\sigma,\sigma') = 0$ if $\sigma \neq \sigma'$. (Apart from a constant term added to H, the Ising model is the case $j = \frac{1}{2}$.) Define the $M(\sigma) = <\sum_x\left[\delta(\sigma_x,\sigma) - \frac{1}{n}\right]>$, that is, the number of sites in state σ minus what it would have been if the system was totally disordered.

C.6   The Ising model on square and cubic lattices

The Ising model on a square lattice was solved exactly by Onsager in a piece of exquisite mathematical physics. The solution is in terms of elliptic functions and uses special tricks that only work for the Ising model in a plane. So we will not pursue it further. The solution is still very useful as a way to test theoretical ideas. Any approach to renormalization must reproduce the exactly known critical exponents when applied to the two dimensional Ising model.

On a cubic lattice, it appears that the model cannot be solved in terms of such well known functions of mathematics. But there are excellent approximation methods (Kleinert and Schulte-Frohlinde, 2001) that can give the critical exponents as accurately as you want. The accurate prediction of the critical exponents of magnets and vapors by the Wilson–Fisher fixed point of the three dimensional Ising model is one of the triumphs of theoretical physics.

C.7   Iterations of a function

To understand the universal aspects of dynamics, we will study the simplest case of iterations of a function of a single real variable. See Chapter 16 of (Rajeev, 2013) to get started on chaos. The original references (Feigenbaum, 1978) are by now classics and are well worth reading.

We will study the iteration of the function[3]

$$f(x) = 1 - ax^2$$

for various parameter values a. It has two fixed points $x_\pm = \frac{\pm\sqrt{4a+1}-1}{2a}$. The value of the derivatives at the fixed points are $f'(x_\pm) = 1 \mp \sqrt{4a+1}$. Thus, for $a < \frac{3}{4}$ one of the fixed points is stable (i.e., $|f'(x_+)| < 1$); most orbits tend to it. As we increase a above $\frac{3}{4}$, both fixed points become unstable. What happens then?

We note that f also has a periodic orbit of period 2 (also called a 2-cycle), $y_\pm = \frac{1\pm\sqrt{4a-3}}{2a}$. You can verify that $f(y_-) = f(y_+)$ and $f(y_+) = f(y_-)$. The derivative of the iterate is

$$\frac{d}{dx}f(f(x)) = f'(f(x))f'(x)$$

so that at the periodic orbit

$$\frac{d}{dx}f(f(x)) = f'(y_-)f'(y_+) = 4 - 4a.$$

As a increases just above $\frac{3}{4}$ this 2-cycle crosses over from being unstable to stable. Thus, most orbits will then tend to this 2-cycle.

As we increase a further, this 2-cycle also becomes unstable, at $a = \frac{5}{4}$. You should repeat the whole process and find that a 4-cycle becomes stable now. There is a sequence of points $A_1, A_2, \cdots$ at which the period of the stable orbit doubles. These values converge to a point A_∞. When $a > A_\infty$, orbits are chaotic, there being no stable periodic orbits. The situation right at A_∞ is most

---

[3] We choose the origin of x so that the function is symmetric around the point $x = 0$ instead of $x = \frac{1}{2}$ as in (Rajeev, 2013).

interesting: it is the critical point at which chaos sets in. The rate of approach to this criticality is exponential $A_n \approx A_\infty + \alpha^{-n}c$. Numerical experiments show that the constant α is universal. That is, α is independent of the choice of function f. Any unimodal function (continuous function with one maximum) gives the same answer for α, although the period doubling points A_n are themselves not universal. This universality is a truly remarkable fact, reminiscent of the universality of thermodynamic systems near a critical point.

To understand this further we first make an approximate calculation first.

C.7.1   Quadratic approximation

The iterate of f is a quartic function

$$f(f(x)) = 1 - a + 2a^2x^2 - a^3x^4.$$

If we rescale the coordinate by a constant $x \mapsto \frac{x}{\alpha}$ and choose α appropriately, we can renormalize it so that its value at the origin is one. Define

$$f_1(x) = \alpha f\left(f\left(\frac{x}{\alpha}\right)\right), \quad \alpha = \frac{1}{f(1)}.$$

Calculating gives

$$f_1(x) = 1 - R(a)x^2 - a^3(1-a)^3x^4, \quad R(a) = 2a^2(a-1).$$

As we iterate further we will get higher and higher order polynomials. So as a first step let us make the approximation of ignoring the quartic term. Then the result of iterating f once and rescaling is to change the parameter of the map by

$$a \to R(a).$$

Thus, the 2^nth iterate of f is approximated by replacing a by the nth iterate of R. To take the limit $n \to \infty$, we now seek the fixed point of R itself:

$$a_* = \frac{1}{2}\left(1 + \sqrt{3}\right) \approx 1.36603.$$

Note that the rescaling is then by

$$\alpha(a_*) = -(1 + \sqrt{3}) \approx -2.73205.$$

The rate of approach to this fixed point is

$$\delta \equiv R'(a_*) = 4 + \sqrt{3} \approx 5.73205.$$

So a_* is an unstable fixed point. We now look for a general theory that goes beyond the quadratic approximation.

C.8   Feigenbaum's renormalization dynamics

We consider the space of all even functions of a real variable, which analytic at the origin:

$$f(x,a) = 1 + a_1 x^2 + a_2 x^4 + a_3 x^6 + \cdots$$

The coefficients (collectively called a) parametrize this set. By a rescaling we can always choose the leading term to be 1:

$$f(0,a) = 1. \tag{C.1}$$

The original single parameter a is now replaced[4] by the infinite sequence $a = (a_1, a_2, a_3 \cdots)$. (In the jargon of renormalization theory, these are the "coupling constants.")

We can think of a step in the iteration as the replacement of f by $f(f(x))$. The effect of the iteration can be captured by the change in the coupling constants:

$$f(x, R(a)) = \alpha(a) f\left(f\left(\frac{x}{\alpha(a)}, a\right), a\right), \quad \alpha(a) = \frac{1}{f(1,a)}.$$

The scaling $\alpha(a)$ will depend on all the parameters, and is chosen so that the normalization condition of eqn (C.1) is satisfied. The function R maps infinite sequences to themselves: it is the *renormalization map*. We have no hope of getting an exact formula for $\alpha(a)$ or $R(a)$. Numerical approximations that keep a finite number of terms will have to suffice.

C.9   The Feigenbaum–Cvitanonic equation

We now seek a fixed point of the map R. That is, an infinite sequence of parameters such that

$$R(a_*) = a_*.$$

This corresponds to a function satisfying the Feigenbaum–Cvitanovic equation

$$g(x) = \alpha g\left(g\left(\frac{x}{\alpha}\right)\right), \quad g(0) = 1; \tag{C.2}$$

for some number $\alpha = \alpha(a_*)$. Near the fixed point,

$$R(a) \approx a_* + R'(a_*)(a - a_*) + \cdots$$

where $R'(a_*)$ is an infinite dimensional matrix (Jacobian) of derivatives of R at the fixed point. In the immediate vicinity of a_* (or g in the space of functions) renormalization amounts to taking powers of $R'(a_*)$. The most important effect will be in the direction corresponding to the largest eigenvalue. Any direction corresponding to an eigenvalue of magnitude less than one will die out after some iterations. (In the jargon of renormalization theory, they are *irrelevant*.)

For the map R we will see that there is exactly one eigenvector of $R'(a_*)$ with eigenvalue greater than one; this eigenvalue δ is called δ. The numbers α, δ (Feigenbaum constants) are universal within the class of functions allowing a polynomial approximation.

---

[4] A minor point of notation is that what we used to call a now corresponds to $-a_1$.

C.9.1   Quartic approximation

For example, keeping only the terms up to order four in x we get

$$\alpha(a) = \frac{1}{1 + a_1 + a_2}.$$

$$R(a_1, a_2) = \left\{ 2a_1 \, (a_1 + a_2 + 1) \, (a_1 + 2a_2) , \; (a_1 + a_2 + 1)^3 \left(a_1^3 + 6a_2 a_1^2 + 2a_2 a_1 + 4a_2^2 \right) \right\}.$$

There is a fixed point at

$$(a_{1*}, a_{2*}) \approx \{-1.52224, \, 0.127613\}.$$

That the value of a_2 and the change in a_1 are indeed small is encouraging: we did not move too far away from the quadratic approximation. The eigenvalues of R' at this point are 4.8442, -0.485562. The largest eigenvalue is the quartic approximation for δ (getting closer to the exact answer, from 5.7 in the previous order). The next largest eigenvalue is less than one in magnitude, indicating that it represents corrections that will die out under iterations: a further encouraging sign.

Also, the value of α at this approximate fixed point is

$$\alpha(a_*) \approx -2.53403,$$

again geting close to the exact answer.

C.9.2   Sixth order approximation

So we keep going to the next order

$$f(x) = 1 + a_1 x^2 + a_2 x^3 + a_3 x^6 + \cdots$$

to get

$$\alpha(a) = \frac{1}{1 + a_1 + a_2 + a_3}.$$

R is now a rational function too complicated to display nicely. We can use the previous fixed point (supplemented with $a_3 = 0$) as the starting point for Newton–Raphson to find a fixed point of R:

$$a_{1*} = -1.52184, \quad a_{2*} = 0.0729316, \quad a_{3*} = 0.0455086\}.$$

Then we can find the eigenvalues of R' at this fixed point to be

$$4.51651, \quad 0.509973, \quad -0.0890154.$$

The largest eigenvalue gives

$$\delta \approx 4.51651.$$

Notice the next eigenvalues get smaller and smaller: indication of the convergence of the procedure.

Moreover, we can calculate α at the fixed point:

$$\alpha \approx -2.47891.$$

It looks too formidable to take this semi-analytic method to higher orders. A purely numerical procedure is needed.

C.9.3 High precision

The best approach (Briggs, 1991) seems to be to solve for the fixed point of the renormalization map

$$g(x) = \frac{1}{g(1)} g\left(g\left(g(1)x\right)\right), \quad g(0) = 1,$$

with a polynomial approximation[5]

$$g(x) = 1 + \sum_{k=1}^{N} a_i x^{2i}$$

with an N of about a 100. If we impose the above equation at some points $x_1, x_2 \cdots x_N$ (e.g., $x_i = \frac{i}{N}$) we get enough conditions to solve for N unknowns a by the Newton–Raphson method. This determines the function g as well as the constant $\alpha = \frac{1}{g(1)}$ to high precision.

The infinitesimal perturbation of the above equation gives a linear operator (equivalent to R' of the last section)

$$L\psi(x) = -\alpha\psi\left(g\left(\frac{x}{\alpha}\right)\right) - \alpha g'\left(g\left(\frac{x}{\alpha}\right)\right) \psi\left(\frac{x}{\alpha}\right).$$

Once g has been determined in the above approximation, an $N \times N$ matrix approximation to L can be obtained again by evaluating at x_i. Its largest eigenvalue is obtained by the power method. (Taking many powers of a matrix kills off all but the largest eigenvalue and corresponding eigenvector.) (Briggs, 1991) gives more decimal places than anyone will ever need:

$$\alpha \approx -2.502907875095892822283902873218215786381,$$
$$\delta \approx 4.669201609102990671853203820466201617258.$$

C.9.4 Experimental verification

Electrical circuits and fluid mechanics provide many examples of chaotic systems that can be studied in a laboratory. It was hoped at first that the period-doubling route to chaos would lead to a general theory of turbulence. But such a theory still does not exist. Particular cases of period

---

[5] In practice, a Chebycheff basis would be better; recall that the angle between x^{2i} and $x^{2(i+1)}$ tends to zero as $i \to \infty$.

doubling chaos has been observed in systems with a small number of effective degrees of freedom. It is possible to measure (Libchaber *et al.*, 1982) the Feigenbaum constants to one or two decimal places, in agreement with theoretical predictions. The Feigenbaum theory plays a role in chaos similar to the Ising model in magnetism or the hydrogen atom in quantum mechanics. It is a simple, exactly solvable, model of chaos that provide a standard of comparison for more elaborate realistic theories. Developing such a theory of turbulence is one of the main challenges facing theoretical physics today.

Exercise C.5 Determine the first few period doubling points for the unimodal map $f(x, a) = a \sin [\pi x]$.

Exercise C.6 Research project Determine the Feigenbaum constants to as high a precision as you can. You can either implement the purely numerical method as in (Briggs, 1991) or a more analytic procedure as above. Use a Mathematica or C program.

Exercise C.7 Research project Identify the stable and unstable directions of the fixed point (eqn C.2) of the Feigenbaum map. Numerically investigate the unstable manifold, for example, plot some typical examples of functions that lie on this manifold.

Bibliography

Aasen, P. O. and Kreiss, G. (2006). *J. Fluid Mech.*, **568**, 451–71.

Aref, H. (1984). *J. Fluid. Mech.*, **143**, 1–21.

Arnold, D. N., Falk, R. S., and Winther, R. (2010). *Bull. Am. Math. Soc.*, **47**, 281–354.

Arnold, V. I. (1966). *Ann. Inst. Poly. Genoble*, **16**, 319–61.

Arnold, V. I. and Khesin, B. A. (1998). *Topological Methods in Hydrodynamics*. Springer.

Aurentz, J. L. and Trefethen, L. M. (2017). *SIAM Rev.*, **59**, 423–46.

Barrow-Green, J. (1996). *Poincare with acute accent and the Three Body Problem*. American Mathematical Society.

Baxter, R. J. (2008). *Exactly Solved Models in Statistical Mechanics*. Dover.

Briggs, K. (1991). *Math. Comput.*, **57**, 435–39.

Buhmann, M. D. (2003). *Radial Basis Functions*. Cambridge University Press.

Burk, F. E. (2007). *A Garden of Integrals*. Mathematical Association of America.

Cassidy, D. C. (1993). *Uncertainty: The Life and Science of Werner Heisenberg*. W. H. Freeman.

Chandrasekhar, S. (1961). *Hydrodynamic and Hydromagnetic Stability*. Clarendon Press, Oxford.

Chandrasekhar, S. (2010). *An Introduction to the Study of Stellar Structure*. Dover.

Chapman, S. and Cowling, T. G. (1970). *The Mathematical Theory of Non-Uniform Gases* (3rd edn). Cambridge University Press.

Chavel, I. (2006). *Riemannian Geometry: A Modern Introduction*. Cambridge University Press.

Choudhuri, A. R. (2010). *Astrophysics for Physicists*. Cambridge University Press.

Cohen, F. R. (2010). *Braids: Introductory Lectures on Braids, Configurations and Their Applications*. World Scientific.

Courant, R. and Hilbert, D. (1962). *Methods of Mathematical Physics, Vol. 2* (1st edn). John Wiley & Sons.

Davidson, P. A. (2004). *Turbulence: An Introduction for Scientists and Engineers*. Oxford University Press.

Dirac, P. A. M. (1938). *Proc. Roy. Soc. Lond. A*, **167**, 148–69.

Do Carmo, M. P. (1976). *Differential Geometry of Curves and Surfaces*. Pearson.

Driscoll, T. A. and Hale, N. (2015). *IMA J. Numer. Analysis*, 1–25.

Faddeev, L. D. and Takhtajan, L. A. (1987). *Hamiltonian Methods in the Theory of Solitons*. Springer.

Falconer, K. (2003). *Fractal Geometry: Mathematical Foundations and Applications* (2nd edn). Wiley.

Feigenbaum, M. (1978). *J. Stat. Phys.*, **19**, 25–52.

Feynman, R. P. (2002). *Feynman Lectures on Physics (Book 2)*. Basic Books.

Fornberg, B. (2017). *SIAM Rev.*, **40**, 423–46.

Fornberg, B. and Flyer, N. (2001). *A Primer on Radial Basis Functions with Applications*. SIAM.

Gerard-Varet, D. and Dormy, E. (2010). *J. Am. Math. Soc.*, **23**, 591–609.

Hasimoto, H. (1972). *J. Fluid. Mech.*, **51**, 471–85.

Hausner, M. and Schwartz, J. T. (1968). *Lie Groups, Lie Algebras*. Gordon and Breach.

Holm, D. D., Marsden, J. E., and Ratiu, T. S. (1998). *Adv. Math.*, **137**, 1–81.

Iserles, A. (1996). *Numerical Analysis of Differential Equations*. Cambridge University Press.

Iserles, A. (2002). *No. AMS*, **49**, 430.

Iserles, A., Munthe-Kaas, H. Z., NÃ¸rsett, S. P., and Zanna, A. (2000). *Acta Numerica*, **9**, 215–365.

Jackiw, R. (2002). *Lectures on Fluid Mechanics*. Springer.

Khesin, B., Lenells, J., Misiolek, G., and Preston, S. C. (2013). *Pure Appl. Math. Q.*, **9**, 291–332.

Khesin, B. A. and Wendt, R. (2009). *The Geometry of Infinite-Dimensional Groups*. Springer.

Kleinert, H. and Schulte-Frohlinde, V. (2001). *Critical Properties of ϕ^4 Theories*. World Scientific.

Kolmogorov, A. N. (1991). *Proc. Roy. Soc. Lond. A*, **434**, 15–17.

Ladyzhenskaya, O. A. (1987). *The Mathematical Theory of Viscous Incompressible Flow* (6th edn). Butterworh-Heinemann.

Lamb, H. (1945). *Hydrodynamics* (6th edn). Dover.

Landau, L. D. and Lifshitz, E. M. (1969). *Fluid Mechanics* (2nd edn). Gordon and Breach.

Landau, L. D. and Lifshitz, E. M. (1977). *Mechanics* (3rd edn). Pergamon Press.

Landau, L. D. and Lifshitz, E. M. (1980). *Statistical Physics* (3rd edn). Butterworth-Heinemann.

Lee, J. M. (1997). *Riemannian Manifolds*. Springer.

Leray, J. (1934). *Acta. Math.*, **63**, 193–248.

Leveque, R. J. (2002). *Finite Volume Methods for Hyperbolic Problems*. Cambridge University Press.

Libchaber, A., Laroche, C., and Fauve, S. (1982). *J. Phys. Lettres*, **43**, 211–16.

Marsden, J. and Weinstein, A. (1983). *Physica D*, 7, 305–23.

McLachlan, R. I. (1993). *Phys. Rev. Lett.*, **71**, 3043.

McNally, C. P. (2011). *Month. Not. Roy. Ast. Soc.*, **413**, L76–L80.

Milnor, J. (1976). *Adv. Math.*, **21**, 293–329.

Mokhasi, P. (2013). Using Mathematica to simulate and visualize fluid flow in a box. *http://blog. wolfram.com/2013/07/09*.

Mumford, D. (1998). *Pattern theory and vision*. Institute Henri Poincare with acute accent, 7–13.

Necas, J., Rizicka, M., and Sverak, V (1996). *Acta. Math.*, **176**, 283–94.

O'Raifeartaigh, L. and Sreedhar, V. V. (2001). *Ann. Phys.*, **293**, 215–27.

Orszag, S. A. (1971). *J. Fluid. Mech.*, **50**, 659–703.

Patankar, S. V. (1980). *Numerical Heat Transfer and Fluid Flow*. Hemisphere.

Poisson, E. and Will, C. (2014). *Gravitation: Newtonian, Post-Newtonian and Relativistic*. Cambridge University Press.

Rajeev, S. G. (2013). *Advanced Mechanics*. Oxford University Press.

Richtmyer, R. D. and Morton, K. W. (1957). *Difference Methods for Initial-Value Problems*. Interscience.

Rosenhead, L. (1988). *Laminar Boundary Layers*. Dover.

Schlitning, H. (1979). *Boundary Layer Theory* (7th edn). McGraw-Hill.

Schwinger, J. (2003). *Selected Papers on Quantum Electrodynamics*. Dover.

Seregin, G. (2015). *Lecture Notes on the Regularity Theory for the Navier-Stokes Equation*. World Scientific.

Serre, J.-P. (1965). *Lie Algebra and Lie Groups*. Benjamin.

Smale, S. (1967). *Bull. Am. Math. Soc.*, **73**, 747–817.

't Hooft, G. (1994). *Under the Spell of the Gauge Principle*. World Scientific.

Tao, T. (2016). *J. Am. Math. Soc.*, **29**, 601–74.

Trefethen, L. M. (2001). *Spectral Methods in MatLab*. SIAM.

Trefethen, L. N., Trefethen, A. E., Reddy, S. C., and Driscoll, T. A. (1993). *Science*, **261**, 578–84.

Tropea, C., Yarin, A. L., and Foss, J. F. (2007). *The Springer Handbook of Experimental Fluid Mechanics*. Springer.

Tsai, T.-P. (1998). *Arch. Rat. Mech. Anal.*, **143**, 29–51.

Varadarajan, V. S. (1984). *Lie Groups, Lie Algebras, and Their Representation*. Springer.

Wendland, H. (2005). *Scattered Data Approximation*. Cambridge University Press.

Weyl, H. (1942). *Ann. Math.*, **43**, 381–407.

Whitham, G. B. (1999). *Linear and Nonlinear Waves*. Wiley-Interscience.

Wolf, G. H. (1969). *Z. Phys.*, **227**, 291–300.

Zachos, C. K., Fairlie, D. B., and Curtright, T. L. (2005). *Quantum Mechanics in Phase Space*. World Scientific.

Index